RESIDENTIAL CONSTRUCTION MANAGEMENT

Managing According to the Project Lifecycle

JOSEPH A. GRIFFIN, PMP

Copyright © 2010 by J. Ross Publishing

ISBN 978-1-60427-022-8

Printed and bound in the U.S.A. Printed on acid-free paper
10 9 8 7 6 5 4 3 2 1

Library of Congress Cataloging-in-Publication Data

Griffin, Joseph A., 1980–
Residential construction management : managing according to the project
    lifecycle / by Joseph A. Griffin.
        p. cm.
Includes index.
ISBN 978-1-60427-022-8 (alk. paper)
1. House construction. 2. Building—Superintendence. I. Title.
TH4811.G75 2010
690'.8068—dc22
                                        2010001971

Direct all inquiries to J. Ross Publishing, Inc., 5765 N. Andrews Way, Fort Lauderdale, FL 33309.

Phone: (954) 727-9333

Fax: (561) 892-0700

Web: www.jrosspub.com

DEDICATION

This book is dedicated to my wife, Melissa. Thank you for your steadfast support.

CONTENTS

PREFACE

The idea for this book began a few years ago as I was seeking a book that would help me improve my ability to manage construction projects. I purchased a few different construction management textbooks from various bookstores and Web sites, and I did find the information helpful. Each of these books covered a number of topics: building materials, vendor management, client relations, project management, contract management, etc. All this information was helpful and useful, but it never seemed to draw the information together so that I could see how it should all be put together to actually produce a finished product. I even purchased books that dealt more specifically with construction project management. These books also were arranged topically with a section for estimating, scheduling, inspecting, and other such topics, but they never did a satisfactory job of bringing this together to show how these processes are integrated. I had a variety of information at my hand, but I could not quite see how to put it together in a way that would allow me to reach the full potential possible. So I kept looking for a solution—a place where I could learn to put all the pieces together.

This led me back to school. I enrolled and graduated from Western Carolina University with a Masters in Project Management. Since my time there, they have started offering a Masters in Construction Management, which is getting great reviews. During my two years attending Western, I continued to work in the construction field and was able to implement and test the lessons being taught. Early on in my studies, I was introduced to the project lifecycle. This was the missing piece from all my previous reading. It is not that the topic of lifecycles is not dealt with in other literature, but it is dealt with as a side note, as a piece that should be included just to make certain that the book is complete. Upon reading more about the concept, it became clear to me that this is how a book should be structured; it should be structured according to the lifecycle of a construction project.

In other books, the topics were arranged topically: scheduling, bidding, planning, etc. I wanted to arrange a book according to the actual order that the work is done, according to the lifecycle. In this way, I start from the time that the construction project is an idea in someone's mind through to the end when the new owner moves into the home. The book is ordered according to the way the actual work of the project is done.

I hope that you, the reader, will find this way of arranging the information to be as helpful for you as it was for me. Areas where my book is helpful, I am most thankful that I have been able to provide that help. In areas where it is unclear and unhelpful, I take full responsibility and welcome your feedback.

I would like to thank J. Ross Publishing for graciously allowing me the opportunity to write this book. They have been wonderful to work with. Thanks Tim. I also would like to thank my tireless personal editor, Leah Wheeler, who was so willing to read and correct all my work. I would also like to thank the professional editors provided by J. Ross Publishing who did an excellent job. Any errors that are still present are where I have changed their work, not where they have changed mine. I also would like to thank the company where I work and my father, the company's founder, for the opportunity to write and for the indispensible experiences that have allowed me to write this book.

My wife, Melissa, has been a tremendous support throughout the writing of this book. She has been my constant encouragement, often reminding me of the end, even when I did not think the end was in sight. She believed in this from the beginning and I am most thankful for her love and support.

ABOUT THE AUTHOR

 **JOSEPH GRIFFIN'S** personal knowledge of construction sites began at the age of 13 when he took his first summer job as a member of a block mason's crew. Throughout high school, he became further acquainted with various aspects of homebuilding as he assisted with laying footings, installing septic tanks, painting, and landscaping at new home sites. After obtaining his bachelor's degree in business management, Joseph went on to begin working full-time in both real estate and the construction industry, while simultaneously earning his master's in business administration through Lenoir-Rhyne University. He later earned a master's in project management through Western Carolina University. He is currently working towards a master's of accountancy at Western Carolina University on a part time basis.

Joseph is a certified Project Management Professional (PMP®), a member of the Project Management Institute, a member of both local REALTOR® associations and the National Association of REALTORS®, and holds the Graduate REALTOR® Institute designation. He is also a member of The Gideons International. He has authored several articles on construction and project management, which have been featured in national publications. Joseph resides in western North Carolina with his wife Melissa, who taught him all that he knows, and their two faithful Labradors. He may be contacted at joegriffin@bengriffin.com.

Free value-added materials available from
*the Download Resource Center at **www.jrosspub.com***

At J. Ross Publishing we are committed to providing today's professional with practical, hands-on tools that enhance the learning experience and give readers an opportunity to apply what they have learned. That is why we offer free an-cillary materials available for download on this book and all participating Web Added Value™ publications. These online resources may include interactive ver-sions of material that appears in the book or supplemental templates, worksheets, models, plans, case studies, proposals, spreadsheets and assessment tools, among other things. Whenever you see the WAV™ symbol in any of our publications it means bonus materials accompany the book and are available from the Web Added Value™ Download Resource Center at www.jrosspub.com.

Downloads for *Residential Construction Management: Managing According to the Project Lifecycle* include a downloadable building specification form, a change order authorization form, a construction schedule, a sample budget, a construc-tion flowchart, a guide to working with bankers for spec home loans, and more.

PROJECT MANAGEMENT AND RESIDENTIAL CONSTRUCTION

Residential construction is something that I have known my entire life. One might even say that it is all that I have known. A few years after my birth, my father founded a construction company. I grew up in that company. My fondest memories are of going to the office with my father on Saturday. This gave me the opportunity to be at his office working with him, and how I enjoyed working with him. I enjoyed it so much as a young child, that through my teenager years, I would work on various construction crews, learning various aspects of the construction process. During college, I would return home in the summers to intern at the company. During this time, I had the enjoyable job of being the office gopher. Whatever took a warm body and very little skill, I was qualified to do. Mainly during this time, I watched people around me and learned how they did their work. I listened to my father sell homes, negotiate contracts, and coordinate construction projects. I learned to review job cards, cost reports, and a host of other skills. I learned the principles of managing a project, even though we never called it project management.

Later on in life, I would return to school to study project management at Western Carolina, and it was there that I was able to put a specific name to all the activities that I had watched being performed on a daily basis for my entire life. The connection between my background in construction and the more formal principles and practices was something of an awakening. My attention was drawn from focusing entirely on constructing the home toward focusing on how one manages the construction process. What I found was that by improving one's ability to manage projects, one could actually improve the homes one built.

This may seem like an obvious conclusion to the reader, and it is. I suppose I was a little dafter than the average guy. It took a little bit for me to see the connection and benefits between a more formal approach to project management and residential construction. But it is this connection that I hope to discuss briefly in this opening chapter as an introduction to the remainder of the book. This explanation is necessary not because most people cannot see how these ideas are related, but because the relationship between these two ideas needs to be brought together in a practical framework that will help improve both the means of managing a project and the quality of the home being built. Therefore, this first chapter will focus on exploring these two topics briefly and will then focus on bringing the ideas together so that the reader will better understand the purpose and the format of this book.

PROJECT MANAGEMENT

Project management is a science or discipline that seeks to plan, monitor, and manage resources in such a way as to bring about the successful completion of a project's stated objectives or goals. The term has become very popular over the last few years, and a clear definition is not always in the forefront of people's minds when they begin discussing a *project* or what *project management* seeks to accomplish. In order to clearly understand what project management is, there are a few ideas or definitions that one should have in mind. This section of the chapter is going to discuss a few of the key concepts of project management, starting with what exactly a project is and is not, and then a couple of other topics will be covered including the triple-constraint and the project lifecycle. The last topic discussed, the project lifecycle, is the basis for the entire book.

Definition of a Project

The term project is a popular one that is bantered about in almost every aspect of life. Someone will say that he is working on a new project at work or that she is working on a new gardening project. The use of the term in titles also has become in vogue in the last few years. There are a variety of project managers, project analysts, project coordinators, and project assistants running through the hallways of numerous businesses. To see just how popular the term project is in today's work environment, one merely needs to run a search on one of the popular job searching Web sites to see how many jobs and positions include the term project in the title or description.

Whenever a term begins to get such broad usage in society, it will typically tend to lose some of its specific meaning. It is certain that if someone were to ask a group of individuals to write out a definition of the word, she would receive a number of answers each having its own nuances. This is not to say that all the definitions would be wrong, but they might not be applicable to every situation where the term might be used.

Because of this possibility of ambiguity that might exist in the mind of the reader, it will be helpful to understand how the term is used in this book. A project can be defined as follows:

> *A project is a temporary endeavor that seeks to create a unique product or service and is constrained by cost, time, and scope.*

Each part of this definition plays a critical role in clearly understanding what exactly a project is, so each part will be briefly considered.

Temporary Endeavor

A project is by its very nature a temporary undertaking. It has a definite start date and a definite end date. During this time, the work that will produce the goal of the project is completed. If a process is long-term in nature, it would not be considered a project by this book's definition. This should not lead the reader to conclude that there are not long-term projects, as there are, but a project is designed to be a temporary undertaking. It will not continue into perpetuity or it is not a project.

A Unique Product or Service

For the sake of this book, the project will typically be undertaken to create a unique product, not a service, as this book deals with constructing buildings. By unique it is meant that the product is one of a kind. If one were to think about it, what is more unique than a project that seeks to construct a home? Even if one is building the same style of home on the same street, it is unique. Even if one were to build the same style of home in the same area, a variety of factors would be unique to each project, such as:

- Land
- Weather
- Subcontractors
- Materials
- Timing
- Client
- Specifications—flooring, colors, etc.

These represent only a few of the items that may be different between the two projects. Because of the unique nature of each project, special planning is required to make certain that the work of the project is done properly and to the specifications provided. Building a home is very different than running a batch of products through a controlled production line, and because of these differences, the work must be managed differently. It is the uniqueness of each project that creates the challenges that the project manager must rise to meet in order to successfully complete the project.

Constrained by Cost, Time, and Scope

Few messages are heard more commonly in the world today than those that speak of the limited resources of the world. No one can listen to the news for any length of time and not hear of how one should endeavor to preserve resources and the environment. Just as the world must grapple on a grand scale with the resources available, so too must the project manager manage the project with the resources available. There are three constraints that are present on every project—cost, time, and the extent of the work. There is a set cost or budget, a set time to complete the work, and a set amount of work to be performed. In every project that is executed, one must grapple with these constraints. The interaction of this dynamic shall be discussed in greater detail later in the chapter, as a key understanding of this dynamic is critical to successfully managing a project.

Now that what is meant by the term project is clear, the triple-constraint that every project faces shall be discussed.

Triple-Constraint

The term triple-constraint was introduced in the last section that defined what a project is. The triple-constraint of a project is cost, time, and scope. The relationship between the factors is depicted graphically in Figure 1.1. The three constraints shown in the figure are those that are used by a variety of project management professionals, but some will use different terms to express the same content. For instance, one might see the triple-constraint referred to as scope, schedule, and resources or cost, time, and quality, or any number of closely related terms. Although the terminology may vary among users, the principle is the same. Every project has limited resources, a limited amount of time, and a limit to the scope or the extent of the work. If one of these factors is altered, it changes the dynamics of the other constraints in either a positive or negative manner. First, each term will be briefly defined, and then the dynamic relationship that exists between these constraints shall be explained through the use of a few examples.

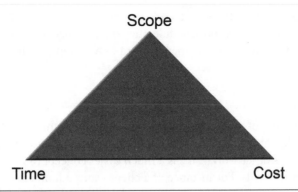

Figure 1.1 Triple-constraint of a project

Cost

Every project has a limited budget. There is a point at which there are no resources remaining to fund the work of the project. If the project manager goes beyond that point, then the work of the project will remain unfinished until new funds are available. Some clients will be more sensitive to cost than others, but everyone has limits. Because of this, the project manager must work with the resources that have been provided. If he is unable to do this, then he will not successfully complete the project.

One of the first steps of beginning a project is making certain that the cost estimate for the project is reasonable and acceptable. If the client has unrealistic expectations about the total cost of a project, then something will need to be modified. An adjustment will be necessary.

Too often cost is seen as referring to money only, but this is not the case. Cost, or the more general term resources, refers to something that is broader than money. But it is not a problem to reduce it to a dollar value. The reason is that if a construction company lacks the expertise to perform a project, given enough money, they can purchase the expertise. This happens all the time through the use of subcontractors, which is popular in residential construction projects.

Time

Time is also one of the critical constraints of every project. A key part of the definition of the project is that it is a temporary endeavor. It has a definite beginning point and a definite ending point by which time the work is to be completed. If the project manager fails to properly manage the project's schedule, he will need to adjust the other constraints in some manner. In construction, the cost of time delays can be quite high. It can result in fines, which is a short-term loss, but it also can have longer-lasting effects. If a construction company develops a reputation for completing projects behind schedule, they might expect to lose some business. Therefore, it is vital that a project manager not only recognizes this constraint, but manages it well.

Scope

Scope has a number of aspects. Primarily, however, it has the idea of a limit to the amount of work that is to be performed. Every project includes some work and by necessity excludes other work. This is a much easier concept to grasp in construction than in other fields. For instance, if a client wants a home to be built according to certain specifications, then it is fairly obvious what the scope of the project is. But in, say, a software development project, the scope of the project can continue to grow and grow much more easily. Remodeling is an area in construction where the scope of a project can be difficult to manage because often clients will want to expand the work beyond what the original project description included. What begins as a bathroom or kitchen remodel soon moves to the living room, or includes adding a sunroom or the like. The point here is not to say that work cannot be added to a project, but it is to say that work cannot be added without affecting the other constraints of the project. One cannot add work without also increasing the cost and time necessary to complete the project.

Another term that is often used in place of scope is quality. It is actually employed later in this book, as in that section of the book it is a more appropriate constraint to consider. The scope by its very definition includes a standard of work or a certain level of quality. If the quality of the project is increased, the cost and the time required to do the work also increases. Therefore, it is important for the reader to be aware of this usage as well.

Now that each of the constraints has been discussed individually, a few examples to explore the dynamics might be helpful.

Cost Example

Imagine that a construction company has taken on a particular building project to construct a home for a client. During the early stages of working through the construction plan, the client learns that the amount she will be able to borrow from the bank is going to be reduced by 20 percent, from $200,000 to $160,000, due to a change in the client's employment situation. This is a fairly large cut in the budget of the project, as a $160,000 home is much different than a $200,000 home in most markets. The client, however, is not deterred and still wants the project manager to construct the same home they were discussing when the budget was $200,000.

If the person reading this book has had any experience in construction and was the project manager in this situation, then she would immediately think one of a few possible thoughts. The first question, however, would probably be can the home proposed be reduced in some manner (square footage, components, etc.) to be within the new budget or will a new house plan need to be located? Because the budget has been reduced, the scope of the project must also change to fit with the new budget. Such a large reduction in budget will typically require a reduction in the square footage or a change in the style of the home or a change in the quality of the components in the home. As can be seen, a change to one aspect will require others to be altered as well.

Time Example

A construction project is underway that is running a couple of weeks behind schedule. When the project manager meets with the client and informs the client about this, the project manager is told that the project better be done in no uncertain terms according to the original agreement. For the sake of the example, the project manager takes this to be a valid threat and moves to take the necessary steps to increase the production speed. In order to do this, bonuses are offered to subcontractors to put this project ahead of other projects on which they might be working. The priority then given to the project by the subcontractors allows the project to be completed on schedule. But in order to accomplish this, the cost constraint was affected, as the cost of the project increased. Hopefully, the project manager's crashing of the project did not also affect the quality constraint of the project, but whenever the speed of the work is rapidly increased, there is always the risk that the quality of the work will suffer as well.

Scope Example

Imagine a remodeling project that consists of replacing the cabinets and floors in a kitchen. As the floors of the kitchen are being taken up, it is discovered that the adjoining living room has hardwoods underneath the carpet. Upon learning this, the client wants the entire home checked to see if hardwoods are present in the other rooms of the home. The search is completed and hardwoods that need to be refinished are found throughout the entire home. The client quickly tells the project manager to include this as part of the remodeling work.

Adding this is not necessarily a problem, and it is doubtful that the client expects this work to be done at no additional cost, but it obviously will affect the cost to the client and the time required to finish the project because of the addition to the scope of the work to be performed. Most of the time, a problem is not created because of such a major change, but a problem is created through a number of small incremental changes that continually adds work to the project. This is known as scope creep and it can wreck a project very quickly if it goes unchecked. Anytime a deviation from the agreed upon work occurs, it will affect the cost of the project and the time required to complete the project, and the project manager needs to both inform the client of this and get approval for the additional work. The point here is that a change to any one of the three constraints will impact the other two constraints in some manner or another.

Because every project shares these constraints, the project manager should keep this concept in the forefront of his mind. Every time a problem arises, the project manager immediately should question how this will affect the triple constraints. If a project manager can have this in mind, then it is more likely that the issues that arise can be addressed properly. This may mean that problems can be corrected, or it may simply mean that the effects are expected and will be coped with to the best of one's ability.

Project Lifecycle

A lifecycle is a progression through various phases of development. The term is used in reference to a number of areas. It is used in reference to biological life, product development, software, and the life of a product after development. The reason that the term is so useful in so many areas is that it helps one think through a process or a phenomenon in a logical manner, which is to say that it allows someone to think through the process or phenomenon in a natural manner.

For instance, if one thinks through biological life in this way, one would think of the organic life form from conception to birth to the passage from childhood to adulthood and on to death. This is called the circle-of-life by some; the more cultic language would be that from dust we came and to dust we shall return. The term lifecycle simply refers to the natural progression of either a process or a natural phenomenon through the natural stages of development.

Projects, like so many other processes, have a lifecycle. There is a natural flow or progression to all projects. The lifecycle of some projects might vary slightly, depending upon the type of project or context of the project, but, in general, most projects can be described as having the same lifecycle.

This book advocates a five phase lifecycle. Although other books and writings may suggest more or fewer phases; the model chosen for this book seems to be well suited for construction projects. The five phases that are presented in this book are not original to the author, but are the names used for the phases in a range of literature. The five phases are:

1. Initiation
2. Planning
3. Executing
4. Controlling
5. Closing

A visual representation of the phases and their relationship to one another is seen in Figure 1.2. The process begins at initiation and then progresses through the other phases and ends at the closing phase. What is also of interest is the fact that the flow of the lifecycle is not entirely linear, but, in some cases, there is an iterative nature to the lifecycle. For instance, during the controlling phase, the project manager may realize that part of the project plan will have to be re-worked in order to accommodate an unknown situation that has arisen, which will require that a portion of the project needs to be re-planned and re-executed. This iterative nature can be seen in Figure 1.3.

The arrows in Figure 1.3 that lead from the bottom of the execution and controlling phase symbols back to the planning phase show that once work is performed and monitored it may need to be re-planned, passing through the planning phase again. Now each phase of the lifecycle shall be briefly discussed.

Initiation Phase

During the initiation phase of the project, the idea for the project is presented and the project is investigated to determine whether or not the proposal should be approved and moved to the planning phase. The idea will come from one of two basic sources in the construction field. It either will come from within the organization or outside the organization; therefore, it will either be an internally-initiated project or an externally-initiated project.

Figure 1.2 Project lifecycle

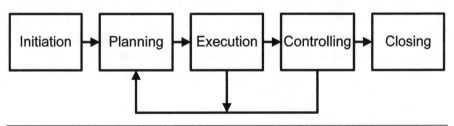

Figure 1.3 The nature of the project lifecycle

An internally-initiated project is one where a member of the construction company suggests it. This might be a new storage building, a new headquarters, or it may be a new spec building that will either be sold or leased. The key point or factor is that the construction company, at the point of initiation, is in some way its own client.

An externally-initiated project is the more traditional type in which a client approaches the organization asking it to perform a construction project. In this type of project, the client is directing the specifics of the project as it relates to what to build and where to build it. Throughout the entire life of the project, the client is seen as the major stakeholder. In this instance, the financing for the project is typically the responsibility of the client.

Regardless of the source of the idea for the project, the primary purpose of the initiation phase is to vet the idea and see whether or not the project should be approved. The type of project will determine the exact process for investigating a project, which shall be discussed later in greater detail; but in general, the various aspects of the project are considered, and based upon that research the project is either approved or declined. If the project is approved, it moves from the initiation phase to the planning phase. If it is declined, it must be forgotten or modified to meet a standard that would allow it to be approved. A major benefit of the initiation phase is that the company has the opportunity to investigate the project before too much is invested. This can help save the company valuable time and money and it helps ensure that projects that are in line with the company's goals are chosen.

Planning Phase

Once the project has moved through the initiation phase and is approved, it goes on to the planning phase. A project manager would have either been assigned to the project during or at the end of the initiation phase. The newly appointed project manager now completes an entire project plan for the construction project. The project plan is the document that both guides the construction project and is used to determine whether or not the project is managed successfully. It is composed of a number of items; a sampling of the major sections are:

1. Project scope and scope management plan
2. Construction schedule
3. Cost control
4. Quality assurance
5. Human resource
6. Communication plan for team members and stakeholders

7. Risk management
8. Purchasing and contract administration
9. Project baselines

Each of these sections is discussed in detail in the following chapters, but as one can see, the project plan is a comprehensive document that seeks to provide guidance to every area of the construction project.

Execution Phase

After the project has been planned, the time arrives for the actual work of the project to be done. In considering the execution of a construction project, there are two aspects to consider. First, there is the actual construction work to consider. The project plan outlines the work necessary to complete the construction project according to the plans and specifications agreed upon by the builder and the client. It is this work that is done during the execution phase. At the end of the execution phase, the client has the home that she requested ready before her. The focus of the execution phase is not only the work of constructing the home, but the focus also includes the work of actually managing the project. The construction plan is without a doubt the most visible plan, but there are other plans that must be executed during the execution phase—the risk management plan, the communication plan, the purchasing plan, etc. These plans must be executed in tandem with the construction plan to complete the project on time and within budget.

Controlling Phase

The controlling phase of the project is concerned with making certain that the work being performed during the execution phase is acceptable. In this sense, the controlling phase is concerned primarily with monitoring both the quality of the work and the quality of the project manager's performance. In order to monitor the work, certain processes and control measures are put in place to both inspect and test project performance. If at any point, the actual work of the project varies from the planned work, which is outlined in the project plan, then steps must be taken to either correct the work being performed or to modify the project plan to match the current realities of the project.

This last point introduces the second aspect of the controlling phase: integrated change control. At times, it becomes necessary to modify the project plan. This typically arises for one of two reasons. First, the client may modify the construction plan in a manner that is conducive to the construction contract. This would be in the form of a construction change order that specifies the work the client would like to have modified. This type of change request typically will require additional time to complete the construction project, and it typically costs the client additional monies. A second aspect of integrated change control concerns how the project plan is actually changed. The project manager will have in place a process for modifying the project plan and alerting those parties who need to be aware of the modification. If this is not done, then problems can arise that can possibly threaten the success of the project.

The controlling phase works in tandem with the execution phase of the project. While the project is being executed, the project is being monitored and controlled. It is as though each process is moving beside the other toward completing the project. Each phase informs and interacts with the other, which means that certain activities are not easily distinguishable as execution processes and controlling processes, but sometimes blur together. The point is that while the work is being executed, it also is being controlled as a means of making certain that the work being executed is the work that was planned.

Closing Phase

After the work of the execution phase has been completed and the work of the project is done, the time has come to close the project. This is typically a relatively short phase as compared to the planning, execution, and controlling phases of the project. In this phase, the project manager is closing out the various aspects of the project to make certain that everything has been completed. For instance, he will inspect work, review payments, close work contracts, ensure the job site is ready for closing, prepare any documents required by the client for the closing transaction on the loan, and prepare performance appraisals for various project stakeholders. The majority of the phase is concerned with completing the various paperwork associated with the project. This means that many builders do not enjoy it, but it is an important phase that should not be neglected.

The project lifecycle is the basis for the organization of the book because it best expresses the natural flow of how work is actually done on a project. The work that is performed during these steps is that which is covered in what is known as the nine project management knowledge areas as defined by the internationally recognized standard, the Project Management Body of Knowledge (PMBOK®). Those nine areas are as follows, and, for the most part, are self-explanatory:

1. Integration
2. Scope
3. Time
4. Cost
5. Quality
6. Human Resource

7. Communication
8. Risk
9. Procurement

These nine areas discussed in the PMBOK® summarize categories for all the work that is done in any project. At different phases of the project's lifecycle these various aspects will get more or less focus. Most books on construction project management focus on these or similar areas without focusing on how they interact during the actual work of the project. In contrast, this book will focus on the project's actual flow of work and show how these and other topics are integrated into the project's lifecycle. This allows the reader to see the direct application and interaction in the life of the project. Now that some of the pertinent areas of project management have been discussed, let the reader's attention be directed to the topic of residential construction.

RESIDENTIAL CONSTRUCTION

Residential construction refers to either the construction or the remodeling of residential dwellings, which I am certain that the reader knows. The field is broad with many specialties. There is much that has been written and could be written about such a broad topic. In most books related to the topic, generally one will find two general types of books. First, there is the book that discusses construction techniques. This type of book will discuss all the specific information that relates to the nuts and bolts of actually building a home. It will discuss different types of footings, different framing techniques. It will provide an overview of plumbing and electrical work. It will discuss the pros and cons of various material types. All this is necessary and beneficial information. Without it, one can hardly manage a construction project, for if they do not understand to some degree the different aspects of construction techniques and materials, then they shall find themselves ill-equipped to manage the construction of a home.

The second general type of book provides guidance on how to manage a construction project. This type of book does not discuss construction techniques or construction materials, instead it discusses construction management. The purpose of this second type of book is to teach someone how to guide the work. It typically will set out to accomplish this by discussing the various types of work that are necessary in order for the actual work of building the home to be possible, as well as the type of work that might go on during the construction of the home. It does this topically. For instance, there will be a chapter on bidding, so that one can learn how to estimate the cost of building the home and submit a bid. There will be an overview of contract law most likely and a chapter on scheduling. Other chapters might cover cost control, job site management, and other similar topics.

These are all necessary topics to cover for someone who wants to be successful in construction management. This book is a little different, however.

Instead of approaching this information topically, this book seeks to tackle construction management in a more linear manner. This is because it is important to know how these processes work together and how to design a custom system to meet one's needs. In order to do this one must understand the nature of projects, as well as the nature of building, and the nature of projects is the project lifecycle. It is the way that projects move from beginning to end. This book takes the project lifecycle, which was discussed earlier in the chapter, and uses it as a skeletal framework for the construction project. Each phase is discussed in detail and the work to be accomplished in each phase is covered. In this way, one not only knows the work that needs to be done, he or she know the order and the types of systems that are necessary to manage the project successfully.

This, I believe, is the benefit of this book. There are books that go into more detail about specific aspects of construction or bidding or scheduling, but I am not aware of one that brings these ideas together and explains them in light of the project lifecycle. This book is not an end-all. There is much more that one needs to know to build a home, but I believe that it will better allow people to manage construction projects because it shows them how to integrate their knowledge into the construction process. With this in mind, we shall turn to the first phase of the project lifecycle: initiation.

REFERENCES

Project Management Institute. 2008. *A Guide to the Project Management Body of Knowledge*, 4th ed. Newtown Square, PA: Project Management Institute.

2

INITIATING THE CONSTRUCTION PROJECT

INTRODUCTION

The initiation phase of a project begins with an idea. In the context of this book, it begins with the idea to build something—a house, a duplex, a small commercial project. The idea could come from an employee, the CEO, the contractor, or the customer. During the initiation phase, the project manager is not concerned so much about the source of the idea as with whether the idea can be transformed into a viable, profitable project. The goal of this phase is to render a judgment on the idea.

In order for the idea to be considered, a process for analyzing it and a standard or a scale against which to measure the results of the analysis must be in place. The process comprises the steps required to gather the necessary information to measure the project according to the standard. The standard is the criteria by which the company decides to take on the project. If the process is faulty, the proposed project will not be accurately measured against the standard. If the standard is too exclusive, projects that could be beneficial to the company will be rejected. If the standard is too inclusive, projects that will harm the long term health of the organization will be accepted.

This chapter will deal primarily with the process whereby the builder gathers the necessary information, evaluates the information, and then determines whether the project is a suitable endeavor for the company. Before the process is outlined, however, it is important to deal with the topics of where ideas for projects come from and how a builder goes about developing the standard or the criteria for deciding whether to accept or reject a project.

Where Do Ideas Come From?

Ideas for construction projects can come from a variety of sources. Broadly speaking, however, they will come from one of two sources. They will come from within the organization itself for the benefit of the company; this type of project is called an *internally initiated* project. Projects can also be initiated externally or by someone other than a member of the organization. These *externally initiated* projects are typically proposed by individuals, businesses, or government or public organizations to meet their specific building needs.

Internally initiated and the externally initiated projects differ in some ways, but they also share a number of steps, which will be detailed in the next major section. The internally initiated project is by its nature a speculative project, as there is no specific purchaser or customer yet known. The company is building the home or office building with the hope that a qualified buyer comes willing, ready, and able to purchase or lease the property. Because the company does not have this buyer in mind, it must choose where to build the home, what style of home to build, how much to price the home, and the like. With an internally initiated project, the construction company takes on much more responsibility and risk. Once the company has accepted this increased risk and chosen the site and the plan, then the steps of the initiation phase are similar to that of the externally initiated project. Below are the steps for each so that the reader may compare and contrast the two types.

Steps in an internally initiated project:

1. Perform market research
   a. Building plan and key feature list
   b. List of acceptable locations
   c. Financial needs analysis
   d. Projected sales price
2. Create a summary of proposed work
3. Preliminary cost estimate
4. List key stakeholders
5. Preliminary risks and rewards analysis
6. Preliminary scope statement

By contrast, the externally initiated project will typically carry a much lower degree of risk, because the customer is initiating the project. If the company properly vets the potential client and takes proper steps during the contracting phase, the company can have a high degree of assurance of being paid for the work performed. Because the client will provide a location, a set of plans, and financing for the construction project, the initiation phase includes fewer steps, as less research is necessary.

Steps in an externally initiated project:

1. Create a summary of proposed work
2. Preliminary cost estimate
3. List key stakeholders
4. Preliminary risks and rewards analysis
5. Preliminary scope statement

When to Assign a Project Manager

Even before beginning the steps for initiating a project, the question of who will lead the project should be considered. The project manager should be assigned to the project as early as possible. The sooner the project manager can assemble the core project team, the better it will be for the project. Some companies may call this individual the construction manager. The title is not as important as the function. The project manager is the individual who will be responsible for planning, executing, controlling, and closing the project when it is completed. Therefore, the sooner this assignment can be made to the project the better.

THE INITIATION PROCESS

As stated above, the purpose of the initiation phase is to perform the research necessary to make an informed decision about whether to proceed with a particular project. The information necessary to make that decision will depend on the type and source of the project. What is needed is a process to move from idea to project. There are four steps to initiate a project, regardless of whether it is internally or externally initiated. They are:

1. Creating the project proposal
2. Reviewing the project proposal
3. Approving the project proposal
4. Transitioning to the planning phase

Creating the Project Proposal

Creating the project proposal requires gathering the information necessary to make a decision. Each company will have their own criteria for selecting a project. The information presented in this chapter may or may not be what a particular company seeks, but this chapter does include the type of information that is necessary to make an informed decision. Internal and external projects require different types of information, which will be covered in the remainder of this section.

Externally Initiated Project Proposal

Because the externally initiated project contains all the basic steps for initiating both types of projects, it shall be considered first, and then the steps that are unique to initiating an internal project will be covered.

The biggest difference between initiating an external project and initiating an internal project is the client. In an internal project, the end-user of the structure is unknown at the point of initiation. For an external project, the client approaches the construction company to initiate the project. The client provides a site location, chooses the plans, usually has a time frame in mind, and sets a budget based on her financial situation. The client must be considered at every point while initiating the project. The steps to initiate an externally initiated project are:

1. Create a summary of proposed work
2. Prepare preliminary cost estimate
3. List key stakeholders
4. Perform preliminary risks and rewards analysis
5. Develop preliminary scope statement

Create a summary of proposed work By reviewing the client's plan, considering the time frame, and analyzing the finances, a summary of proposed work (SOPW) is developed for approval. Once the project manager writes a SOPW, reviewing that document with the potential client to ensure agreement on all expectations is helpful. In the SOPW for an external client, the project manager may want to include additional information, such as a specific set of plans and building specifications as provided by the client. As the SOPW does not contain within its narrative form an exhaustive description of the proposed structure, it is important to reference both standards of workmanship and materials. As well, a full set of blueprints, which would include the layout, elevation take-offs based on the actual lot, cabinet diagrams, and electrical, HVAC, and plumbing plans, should be referenced. The more exhaustive the supporting documentation, the better protection there is for the construction company. If the client will be providing any services or materials as it relates to the construction of the building, it is important that those items are identified as well.

Prepare a preliminary cost estimate Creating a preliminary cost estimate is an important step. I will cite three basic methods one can choose from to create this preliminary cost estimate. This is by no means an exhaustive discussion, but it provides the reader with a starting point. Kerzner identifies these three major types of estimates as follows (Kerzner 2003, 514):

1. Order-of-magnitude or ballpark
2. Approximate or top-down
3. Definitive or bottom-up

The ballpark estimate is the quickest and easiest to perform. Typically, it is based on some type of multiplier such as price per square foot. For instance, if a company has been building homes for around $150 per square foot, and someone requests a quote for a 2000 square foot home, the quote would be $300,000. This type of estimate is obviously easy to provide if a price per square foot multiplier is at hand, but the methodology is not overly reliable. It has a plus-or-minus 35-percent margin of variation, which means that it is likely that the actual cost to construct the home could range from as low as $195,000 to as high as $405,000. If the client were requesting a quote on a home that is more utilitarian in its design because of the client's large family and limited budget, a no-frills 2000 square foot home could be built for less than $200,000. But if the client were seeking a rather palatial-style home with numerous upgrades and extras, the home could easily cost even more than $405,000 to build. Thus, providing a ballpark estimate on the spot without reviewing building plans and specifications is not a particularly reliable way to quote potential projects. Those builders who participate in more utilitarian projects will find this method to be more reliable than those participating in more customized projects.

The second cost estimate method is the approximate or the top-down estimate, which is based on more concrete, historical data, such as past similar projects. This method of estimating is statistically more reliable than the ballpark estimate as it typically has a plus-or-minus 15-percent margin of variation. This method of preparing an estimate depends on past projects that are similar in scope and type. For instance, if a builder were asked to build a certain style of home that he had built a year or so previously, the quote could be prepared by going back and reviewing the previous project and updating the estimates based on current pricing. For instance, if the lumber cost on the previous project was $15,000 and prices for lumber have increased by 20 percent, an updated estimate of $18,000 for the current project would be reasonable. Other costs may have slightly decreased; therefore, particular portions of the estimate may be lowered. This type of estimate is more time consuming than the ballpark estimate, but it is also more reliable and realistic and therefore much preferred. For builders who repeatedly build similar styles of homes, this may prove to be a very usable method. It will not be as useful for highly custom builders.

The final method is the definitive or the bottom-up estimate. This method is both the most time consuming and the most reliable estimate. It is most reliable because of the amount of time spent reviewing plans and specifications and requesting quotes from vendors or subject matter experts. With this method of estimating, the project manager will look at each piece of work required to build the home and provide a detailed quote. Each individual quote will be added up to a total cost. The amount of detail necessary depends on the desire for accuracy. The size, scope, and financial constraints of the project will determine whether such a methodology is appropriate for a given project. This is a very accurate method and typically only has a variation of 5 percent or less.

Once the project manager has selected the method that seems to be the most appropriate to the situation, he should consider additional costs, which are not directly related to the construction of the home. These are generally categorized as overhead costs. Overhead cost refers to costs that are not directly assignable to a portion of the project, such as the salaries of employees who work on the project in their normal capacity. This might include the project manager, accounts payable, legal counsel, or a host of other individuals or groups indirectly associated with the project in some capacity. Typically, the company will have an overhead rate that is applied to the project. The company's accountant will help determine what the amount should be.

List key stakeholders Stakeholders are individuals who have a vested interest in the project. Some people have minor interests in the project, while others may have a key interest in the projects. Most stakeholders will fall somewhere in the middle. For an internally initiated project, the most important stakeholder is the company. The individual within the company who has the authority to authorize or cancel the project represents the interests of the company. Many times this individual is using the information provided in the previous section to help determine whether the proposed project should be undertaken. Listing key stakeholders and their relationship to the project provides critical information for the key decision makers. An example of this is as follows.

The company for which I work was seeking land in a certain county to expand the company's presence to new markets through offering an affordable housing option. Upon identifying a vacant property that was suitable for an affordable housing development, negotiations were begun with the property owner and key stakeholders. During the initiation phase, the director of planning from the governing municipality alerted us that he would not approve a property for development unless certain design standards were met. The requirement did not concern the quality of the work or the materials, but the exterior appearance of the property. The cost of implementing his design requirements would have been prohibitive for an affordable housing development, leading to the termination of the project during the initiation phase. For many projects, the local planning department is a stakeholder, albeit a minor one. By researching potential stakeholders during the initiation phase of the project, the builder can discover whether a certain stakeholder's stance on the project could lead to the failure of the project. Therefore, it is helpful to classify and list stakeholders, as well as any concerns that may exist about each stakeholder. This takes a little more effort on the part of the project manager, but it could be the difference between project success or project failure.

Stakeholders can be ranked according to their potential influence on the project. Below are some sample classifications:

1. Key stakeholder: The key stakeholder wields considerable influence over the project. This could be the overseeing decision maker from the company, who is called the project sponsor. It could also be people who elevate themselves to this level because of their opposition to the project. For instance, construction of an apartment complex in a mostly residential neighborhood may engender protests from neighborhood residents. If these residents are organized and determined, their opposition could become an obstacle to the success of the project. A suitable response must be developed to manage such an obstacle. Other key stakeholders would include the project manager, the core project team, the financing institution, and any other individuals and organizations that could have a major impact on the project.

2. General stakeholder: General stakeholders are people who will have an interest in the project, but their influence will not greatly impact the direction of the project. Most of the construction subcontractors, the local government officials, and others will fall into this category. Not all the general stakeholders will be known at this phase of the project. Therefore, importance is placed on identifying all known key stakeholders.

Stakeholder management plays an important part in the success of any project, but especially construction projects because so many people are involved at various points of the project. It is such an important part of the project that a portion of the project plan will focus on stakeholder management. During the initiation phase, however, it is most important to simply identify and list any concerns or cautions associated with each key stakeholder. A more detailed approach to managing all stakeholders will be developed later on during the planning phase.

Perform preliminary risks and rewards analysis The construction industry is abundant with opportunities to reap profitable and personally fulfilling rewards, but it is also vulnerable to loss and failure. When considering a potential project, it can be very helpful to list not only what the benefits of a project are, but also what the potential pitfalls are. By carefully weighing the potential risks and rewards, the builder can help ensure project success.

Risks include anything that could reasonably go wrong with the project. There are some risk events that are likely and some that are unlikely. For instance, if a house is being built in Kansas, it is unlikely that a tsunami in the Pacific Ocean will destroy the property, but damage from a tornado may be a strong possibility. Therefore, this list should contain those risk events that are reasonable concerns for this particular project. Questions to ask are:

- Does it appear that the project will be profitable?
- Do we have the expertise necessary to build this project?
- Can we acquire the expertise in a reasonable time for a reasonable price?
- If this project fails, does it threaten to bankrupt the company?
- Will this project further the company's long range objectives?

Attempting to answer these and similar questions assures that the project manager is entering the project with eyes wide open. Failing to do this or a similar type of assessment creates a higher probability of the failure not only of one project but of the entire company. It is critical that a risk assessment such as this be performed early on.

The next aspect of the project to consider is much more enjoyable to think about—the potential benefits. Benefits, like risks, come in all forms. There are monetary benefits, organizational benefits, goodwill benefits, personal enrichment benefits, as well as others. All that could go wrong has already been listed; the question now is what could go right. If the project is a success, what are the benefits that the company can expect to enjoy by virtue of its success? Some of the questions to ask are:

- Will this project increase the company's profitability?
- By doing this project, will the company be able to hone or add a new skill set?
- Will this project lead to more projects?
- Will this project allow the company to enter a new market that was unreached before?

Although some of these questions may not seem directly applicable to only this project, they are applicable to the long-term success of the company. Projects should be strategically chosen to build and enhance the long-term success of the company. If a company's only drive is short-term profitability, then there is a high probability that the company will fail in the long term. By strategically analyzing projects for more than their profitability, the company can develop its ability to compete in the marketplace in a manner that will help ensure long-term success.

In the case of an externally initiated project, the potential client is the initiator of the project, and that client is responsible for financing the work. Therefore, it is also important to research his financial ability early on and determine whether he has the financial resources to see the project through. If he is acquiring a construction loan from a financial institution, then the lending institution should be able to provide a loan preapproval letter. Preapproval letters do not guarantee that the potential client will be able to secure a loan, so the construction company may have to do additional research. Asking the potential client to provide a credit report, as well as bank statements may seem rather intrusive, but the client is asking the company to place themselves in a situation that requires considerable risk. Given this fact, if the client is unwilling to prove his ability to satisfy the financial obligations, this could be an indication that he in fact does not have adequate resources.

Develop preliminary scope statement The preliminary scope statement should be seen as a guiding statement for the project. It sets the course to follow and the boundaries to stay within. This is why it is wise for the scope statement not only to include what work is to be included in the project, but also what work will be excluded from the project.

The importance of a scope statement will vary between projects. For instance, the scope statement for a newly constructed home which references a specific set of construction plans will be rather clear to anyone who can read the plans. Although the scope statement is important to anyone who wants to get a general feel for the project, its role is not very major.

Anyone who has ever done remodeling work is probably familiar with how the project can continue to grow and grow and grow. This can create major conflicts and problems between the project manager and the client. In this instance, the scope statement plays an important role as it should specify the limit or the boundaries of the work. For instance, if the client wants a kitchen remodeled, then there is no reason to be working in the master bathroom. If the project grows to include the bathroom, but new quotes and contracts have not been executed, then the project has experienced scope creep. If this happens, what contractual guarantees does the project manager have of compensation for the work? Will the courts determine that this work was in accordance with the original construction contract? Well, if the scope statement limits remodeling work to the kitchen, then it should be clear to anyone who reads it that the bathroom was additional work, for which payment must be made.

Therefore, the preliminary scope statement should be a general statement about the work that is being proposed, as well as the boundaries to that work. If the project progresses from the initiation phase to the planning phase, the scope statement will need to be revisited and updated to include the work that has been agreed on in the construction documentation. Therefore, do not think of this scope statement now as the final draft, but a good approximation of all the work that the project is to include. Typically, it is good for it to reference a set of plans and drawings associated with the work.

These previous five sections comprise the steps to initiating an external project; they are the same steps a builder would perform when analyzing an internally initiated project as well. The one exception with the internally initiated project is that the company needs to perform some market research to determine what type of home to build and where to build it, which is no mean task. Therefore, the process will be covered in the following section on the internally initiated project.

Internally Initiated Project Proposal

No specific client or customer is typically known for internally initiated projects. Because of the speculative nature of the project, it requires an additional step during the initiation phase of the project known as market research. Other than the market research step, which is performed first, the steps of an internally initiated project are the same as for the externally initiated project.

Perform market research Performing market research is a vital step to initiating a successful internally initiated project. The market research will indicate whether it is likely that the project idea will contribute to the long-term success and financial welfare of the company. Doing the market research can be time consuming, but it is a wise investment of time and resources. Consider this example:

If someone within the company were to have the idea to build a home, the following tasks would need to be addressed:

- Locate a building site
- Choose a style of home to build and what features to incorporate
- Calculate expected cost of construction
- Calculate expected cost of marketing and selling the property
- Calculate estimated sales price of home
- Calculate estimated return on investment (ROI)

Whether the builder is considering a light commercial property or a residential property, the basic research would be the same. Armed with this information, the company will be able to make an informed decision about whether it would be wise to speculate on such a project.

If the builder wanted to construct a light commercial property or a residential rental property, the research would refocus on gathering the following information:

- Background on local rental market
- Site investigation (proximity to conveniences, traffic count, land cost, zoning regulations, etc.)
- Expected cost of construction
- Cost of long-term financing
- Expected rental income
- Expected cost of maintaining and managing the property(ies)
- Estimated ROI

By gathering this information, the builder would be able to make an informed decision about constructing either a light commercial or residential rental property. Failing to do this type of research during the initiation phase of the project risks failure of the project, as well as failure of the company, if the company is unable to sustain the loss.

To build a spec home, the following tasks must be accomplished:

- Locate a building site
- Choose style of home to build and what features to incorporate
- Calculate expected cost of construction
- Calculate expected cost of marketing and selling the property
- Calculate estimated sales price of home
- Calculate estimated ROI

Locate a building site Locating the building site is first on the list because the location of the site needs to be known before the builder can gather the information in later steps. As the saying goes, the three most important things in real estate are location, location, and location. Overcoming a poor location is very difficult. The best place to start to research location is with a real estate professional. A REALTOR® has access to a vast amount of market research. The type of information that the builder needs to make an informed decision about location is:

- Current home sales trends: Where are people moving from? Where are people moving to? What neighborhoods are popular?
- Current lot sales trends: What lots or parcels or tracts of land have recently been sold and purchased? Where are residential and commercial projects currently active in the area? What size of lot is most popular? What lot features are most popular (proximity to walking parks, greenways,

sidewalks, water frontage, long-range views, etc.)? What is the average cost of a lot by area?

- Current new construction trends: Are current building permit applications located in a certain area of town or of the county?

The type of information listed above will reveal where people want to live, what type of lots they are seeking, and where the competition is located. All this information can be gathered fairly easily. A competent real estate professional can provide the statistics on home and lot sales and lot features. The local planning department can provide information on current development, as well as those developments that are in the pipeline. The local building inspection department will be able to provide information about how many building permits have been filed and for what areas. All this is public information that anyone can access.

While doing this type of research and visiting the neighborhoods, the builder or project manager will be able to start considering the size and style of home that is going to be built. More detailed information will be provided later, but for now it is important to note the general price ranges of homes that are being built on the lots that are being considered. For instance, if the builder is leaning toward lots in a neighborhood that consists primarily of $500,000 homes, she or the project manager will need to know that she will be able to afford to build such a home. If she can afford to build only a $250,000 home, there is little need to research neighborhoods that are filled with $500,000 homes.

Each neighborhood most likely will have a set of building restrictions attached to the deed or subdivision plat map, or it will have an architectural review board composed of members of the home owner's association. As soon as the builder or project manager begins to seriously consider a lot in a specific neighborhood, she needs to meet with the architectural review board or get a copy of the building restrictions, as well as consult with local zoning, planning, and building inspection officials to learn what the construction expectations are. The sooner she gathers this information, the sooner she will be able to proceed to choosing a plan and features and completing an initial cost estimate.

Once the research has been done and she has an idea of what lot to purchase, how does she make certain that another buyer does not come along and purchase the lot? There are a couple of options. A possible solution would be to simply go ahead and purchase the lot. This would obviously remove the risk of someone else purchasing the lot. But if later research reveals that the site would not be conducive to the goals of the project, the purchaser is now the owner of a lot that is not wanted. Another possible solution is to purchase an option on the property. For instance, purchasing a 30-, 60-, 90-, or 180-day option to purchase the property would allow ample time to perform the necessary research with the only monetary risk being the cost of the option. The cost of an option is not a set amount, as it typically varies based on the time requested and the market value of the lot. But this is a much less expensive option than going ahead and purchasing a lot that may not actually be used.

Choose a home style and features After researching possible locations and finally locating a suitable property, the project manager will most likely have a general idea of the size and style of home to build, as well as the features that should be incorporated. The size and style of the home will be strongly influenced by the neighborhood chosen for the new home's location.

When making plans to build in a residential neighborhood, the builder needs to get information from the following sources:

1. Deed restrictions or restrictive covenants of the neighborhood
2. Zoning guidelines from the local planning department
3. Style restrictions and building guidelines from the architectural review board of the neighborhood

With the applicable information from these three sources, the builder will have a fairly clear idea of what can and cannot be built on the property. From this point, he can look at other sources of information, such as current trends from the National Association of REALTORS® and the National Association of Home Builders. Each year these organizations perform extensive research on the features that different types of buyers are seeking.

The style and features that the builder is seeking to incorporate will vary depending on the type of buyer. Someone who is building in a resort market will focus more attention on certain features and popular building practices than someone who is building affordable housing. Once the typical buyer is known, the builder can incorporate those features that will have the highest level of appeal.

Calculate expected cost of construction Once the site, the style, and the desired features have been chosen, the builder or project manager is ready to complete an estimated cost of construction. Creating a preliminary cost estimate is an important step. Previously in this chapter, three general methods of preparing a cost estimate were presented. They were the (1) order-of-magnitude method, (2) approximate estimate, and (3) definitive estimate. The reader can review the earlier section of the chapter.

Calculate expected cost of marketing and selling the property The expected cost of marketing and selling the property will vary depending on the local market, as the cost for various forms of advertising varies from market to market and according to the method used. If the construction company has its own marketing and advertising division, the cost will most likely be assigned as a percentage of overhead. For instance, if a firm spends $50,000 per year on marketing its properties and builds 50 homes per year, then the firm could assign a marketing cost of $1,000 per home. This, however, may or may not include all the costs associated with selling the home. If a commission has to be paid to a selling agent, the firm would have to consider that commission as well. The firm for which I work has its own in-house selling and marketing staff. In this situation, a fixed overhead amount is applied to each home, although some homes may take more time and expense to sell than others. In firms that do not have their own marketing and sales department, the builder's best choice will most likely be to hire a professional real estate firm to market and sell the property. By choosing an established firm that has a strong track record in new construction, the builder can minimize the marketing time and cost of selling the property. Although there is no set price that real estate firms charge, the average for most markets is between 5 and 8 percent. Regardless of the method chosen to sell the property, marketing costs must be estimated.

Calculate estimated sales price of home The company has three basic choices for calculating the estimated sales price: The company could choose to (1) prepare an in-house estimated value, (2) commission a comparative market analysis from a real estate professional, or (3) hire a professional appraiser. The in-house estimated value should rely on historical sales data, not on a calculation of estimated cost of construction plus desired profit. The project manager should look at the last few months of the sales activity of similar types of homes in similar types of neighborhoods. By reviewing these comparable properties, he can see what values to expect.

The second option is to provide a real estate agent with the information you have previously collected and ask her to perform a comparative market analysis (CMA). A CMA is a relatively straightforward procedure in which the agent will look at comparable sales over the past few months in order to arrive at an estimated value. One benefit of using an agent is that she will be able to provide you with detailed sales statistics for the various properties that have sold. For instance, she will be able to tell you the average days on the market for properties by price range and area. She should also be able to provide insight into which construction features might have contributed to a given home having a higher sales price with fewer days on the market. The agent will also be able to tell you how many concessions sellers were willing to give in order to reach an agreement with the buyer. Whether using an agent or not to actually prepare the CMA, the project manager would do well to seek out an agent to gain access to helpful sales statistics.

The third option is to hire a licensed appraiser to perform an appraisal on the property. This is the most expensive option, but it is also the most informative and will yield the most detailed report and most reliable information. This option should be seriously considered by builders with less experience in estimating the sales price.

Calculate estimated profit Once the previous information has been gathered, then the builder is ready to project what type of profit margin she might expect to earn. A common mistake when calculating estimated profit is to only consider costs that are directly related to the project. When this happens, the project income statement might look similar to what is shown in Table 2.1.

In Table 2.1, the costs directly related to constructing this building are $175,000, which, when subtracted from the estimated sales price, leaves an estimated profit of $25,000. Someone might look at this and believe a 10 percent profit margin to be an acceptable amount, and it might be. But it is doubtful that the builder who undertook this project would actually realize a full 10 percent profit margin; most likely the profit margin will be much less, because only direct costs are being considered. The builder must also consider the indirect costs associated with constructing this home. There is one indirect cost that will almost entirely eliminate the profit margin on the project: the sales commission. Sales commissions vary from region to region, but a 5 to 8 percent commission rate is typical. In this case, there would be almost no profit left. Once the builder begins to consider the other indirect costs associated with constructing this project, most likely he would lose money.

The project manager must count all the costs—direct and indirect. Direct costs are easy to identify; the indirect costs are more difficult to identify. Generally, there are some non-raw material costs that are indirectly related to the construction project. These costs are generally identified as sales and administrative costs.

Table 2.1 Project income statement

Est. sales price	$200,000
Labor costs	($75,000)
Material costs	($100,000)
Est. profit	$25,000

The company's sales staff and administrative team work a certain percentage of their time on this project along with other company projects. Therefore, the project manager must estimate how much time is spent on this particular project, which will allow her to develop a cost estimate for the time and expense. The other type of indirect costs that is almost impossible to link to one specific project is general overhead costs: executive salaries, office equipment, office rent, etc. These are costs to the company, but they are not directly linked to a specific project. Therefore, most companies develop some type of overhead rate to apply to each construction project.

The methods for applying overhead can be rather complex, but a very simple methodology of estimating total annual overhead and dividing it by the total number of jobs completed is one approach. For example, let's assume that a small company expects to construct 10 homes during the current calendar year and the company's general operating expenses for that year will total roughly $250,000. This number includes salaries for employees and support staff, office rent, utilities, and other overhead items. In this example, the company would need to assign approximately $25,000 of overhead expense to each of the 10 construction projects. In the example provided initially, in which the project expected to yield an estimated profit of $25,000 without taking into account any indirect costs, the company would really only be covering their costs by taking on the project, which may or may not be acceptable depending on a number of factors.

The method proposed is not meant to be a guide to calculating true costs or how to accurately calculate an overhead rate, but simply as a reminder that these costs need to be taken into account. If a builder fails to consider these costs, the company is unlikely to prosper in the long term. By taking the time to consider these types of issues during the initiation phase of the project, the builder will learn whether a given project will actually meet the profit expectations of the company.

These steps are unique to an internally initiated project. After the market research has been performed, the project manager can move through the initiation steps discussed previously in the externally initiated project step:

1. Create a summary of proposed work
2. Preliminary cost estimate
3. List key stakeholders
4. Preliminary risks and rewards analysis

After moving through these steps, a report should be prepared that will enable the management of the company to determine whether to proceed with the proposed project.

Reviewing the Project Proposal

Once the project manager has completed the above steps and gathered the information into an appropriate report format, the report is ready to be reviewed. Reviewing the project proposal is no mean task. The reviewer must be able to analyze the report to determine, first, whether the report is accurate and, second, whether the proposed project fits the goals of the company.

Reviewing the proposal for accuracy does not necessarily mean that the one performing the review must rework all the numbers, but it does mean that he should look at the report with a critical eye. He must determine if the report uses accurate facts, and he must determine if the conclusions drawn from those facts logically flow from the facts. For instance, if the report were to say that a particular neighborhood has an excellent sales history with homes priced between $150,000 to $175,000, and the attached sales reports in fact show that the homes priced between $200,000 and $225,000 sell the best, then the author of the report has obviously made an error in creating the report. Although such an error is most likely unintentional, it could lead to a costly decision if not caught. This is simply an error of fact.

The author of the report could have also made a faulty conclusion from the facts. The author may have misinterpreted building and buying patterns and chosen a neighborhood in which homes are difficult to sell. Considering interpretations and analysis is one of the aspects of reviewing the proposal. This implies that the one who reviews this proposal realizes what her role is and what is expected of her. It also necessitates that the person performing this review possess the ability to review the data in the manner discussed above.

If the facts presented are accurate and the conclusions drawn in the report are credible, then the reviewer moves on to consider whether the project meets the goals of the organization. A project that is outside the area of expertise of the company or does not help the company achieve its strategic goals can create numerous problems. If the project is beyond the company's abilities, it will most likely fail to achieve project success, which can lead to lawsuits, disgruntled employees, and an irate client. If the project is not in line with the company's strategic objectives, the person who approved it may put himself at risk of being fired or demoted within the organization.

In order for these pitfalls to be avoided, the reviewer needs some type of a standard or baseline to compare the project against. This ensures that each project is considered according to the same factors and that the organizational competencies and goals are considered during the review process. This can be achieved through a number of means or methods.

The organization may develop a few different types of checklists to guide the reviewer. One such checklist might be a completeness checklist, which lists the information that is to be included in the report. This way the one who is preparing the document and the one who is reviewing the document have a shared knowledge of what should be included.

Another checklist or guide that the reviewer might use is one that contains a list of organizational competencies and objectives against which to compare the project proposal. This list would include things such as desired project sizes, desired profitability, desired types of projects, and the like. If the project does not include those items in the list, the reviewer would make an exception note, which would have to be addressed either by the author of the report or by someone who has the authority to do so, who can override the normal process, such as the president of the company. This type of guide will help ensure that projects are considered on some type of standard basis, which should help improve project selection, which will help the company's project success rate.

In addition to these checklists and baselines, the reviewer must rely on her general feel of the project. Not every project can be reduced to a set of statistics and reports. The reviewer should possess the experience necessary to look at a project, consider all the data, and develop a conclusion based on her experience and expert judgment, which, if proven reliable, should weigh heavily on whether the project should be accepted.

Once the review has been completed by the person within the company responsible for such matters, the reviewer should offer a recommendation. If the proposal is rejected, then the reasons should be noted. If the proposal is recommended, then the proposal will progress to the next step: approval.

Approving the Project Proposal

Approving the proposed project is an exciting step. Much work has been put into developing the proposal, and the work has paid off, as the proposal has been reviewed and given a positive review for approval. The steps involved in approving a project will vary from company to company, but these are the types of considerations one should have during this portion of the initiation phase of the project.

Depending on the type of project, the process for approving a project will vary. The most simple approval process is typically associated with the internally initiated project. With an internally initiated project, there is not a contract to prepare or terms to review or the like. Instead, the company officer who has the authority to approve such projects will typically write some type of letter that grants the project manager the official right to begin the project. This document acts as the project manager's authority to create a project team, begin planning, acquire resources, and all the other steps associated with beginning the project. Many times, a project sponsor will be assigned to the project; this may be the person who originally came up with the idea if that person possesses the necessary authority to act in this capacity. This person is responsible for high-level supervision of the project; his level of involvement will depend on his desire for involvement and on company policy. Once the proposal has been approved, the project manager will gather all the associated plans and documents and begin transitioning to the planning phase of the project, which is discussed in the next chapter.

If, however, the project is an externally initiated project, the approval process will most likely be a little more involved because there are competing parties involved. If the proposal is recommended, the builder must outline terms of the contract for review by the client.*

Most of the time, the company will have a standard contract form. The following is a list of issues that are of primary concern to both the project manager and the client and should be detailed in the contract:

- Construction specifications
- Contract price and construction budget
- Quality standards
- Construction time frame
- Special clauses or notes
- Inspection guidelines
- Builder warranty
- Conflict resolution

Construction Specifications

When the client came to the company to propose the project, she provided a list of construction specifications and a set of blueprints. This may have been augmented during the first half of the initiation phase, and those additions must be considered as well. The final draft of the work being proposed will need to be included as part of the construction contract. The items listed need to be as detailed as possible. For instance, if the home includes a finished driveway, the specifications should list how it will be finished, what size it will be, and the like. The more specific the information can be, the clearer understanding everyone will share about the goals of the project. The project will also progress through the planning phase more quickly, as the management team will not need to gather the information.

---

*Note that I am not an attorney so the recommendations offered here do not constitute legal advice but are merely suggestions and considerations based on my experience.

Contract Price and Construction Budget

A second concern included in the contract is the price and the construction budget. Depending on the type of contract being executed, these may be different numbers. For instance, if the contract is a fixed-price contract, the contract price will be higher than the construction budget, as the difference represents the profit that the company expects to realize for performing the project. If, however, the project is a type of cost-plus contract, the contract price represents the best guess at what the actual construction cost will be, based on the preliminary cost estimate and budget developed earlier in the phase.

This section should also include a payment schedule for the work completed. Typically, there will be a nonrefundable down payment and then draws on the construction loan at various points in construction. If the buyer has a construction loan, then this schedule is typically set by the bank, and the inspections are performed by the bank's agent, not the client. This is an excellent benefit to the contractor, as it makes it easier to secure payment for work performed.

Quality Standards

It is also important to reference a quality standard to which the project shall adhere. Many companies have a construction specification list that addresses the construction techniques and materials. This avoids the customer's complaining of ignorance that a certain material was going to be used in construction. The contract should also reference some type of building standard against which the construction could be compared. This could be the local building code, or it could be a standard set by an independent organization. Many building organizations, such as the National Association of Home Builders are providing building standards related to "green" building, which can be referenced in the contract. Before agreeing to any type of standard, the project manager should make certain that he understands the implications.

Construction Time Frame

The clause that deals with the time allowed to complete the construction project is of utmost importance to the project manager. If an unrealistic amount of time is allotted, the client should expect some disappointment. It is far better to ask for a little extra time and then complete the project early. The clause dealing with the time frame will typically also include extensions that become necessary due to client change orders, natural disasters, weather delays, material shortages, and other unforeseeable events. It will also typically include a per-day fine for going past the target date, which will be charged to the contractor. The builder will desire that this be as low as possible. The actual rate will vary depending on project size and acceptable practice. Typically, it will be enough to make certain that the builder will work her best to avoid it.

Special Clauses or Notes

It seems that in almost every contract there is a collection of special clauses or notes that does not naturally fall into any other section of the contract. Recently, I had to include a guarantee that all drywall used in the home must be manufactured in the United States. It has always been the practice of my company to use drywall manufactured domestically, but the client wanted extra assurance.

The builder should not draft these types of clauses or allow the client to do so. The attorney for either party should both draft and review any special clauses as a means of mitigating misunderstandings and disputes that might arise from these unique clauses.

Inspection Guidelines

The section on inspection guidelines should detail the types of inspections the client can perform at various points of construction. If the client is allowed to enter the job site at will and inspect work, the project manager will have to address two concerns. First, he will have to deal with liability issues. What if the client comes to the job site with friends or family, and someone is hurt? This could create liability issues that could have been avoided if a written inspection and job site visiting policy were in place. Second, whenever a client comes to see work, she will typically see the rough product instead of the finished product. This will lead to numerous calls and questions as to why the work has been done in such a way or whether the work will be left as is, and so on. These types of calls are unnecessary and can lead to wasted time managing issues that need not be addressed. By implementing clear inspection guidelines as part of the construction contract, both parties have clear expectations and understandings, and it can help avoid the situations previously mentioned.

Builder Warranty

In most states, the law requires that the builder provide a 1-year warranty on workmanship and craftsmanship. The question that is of primary consideration with the warranty is whether the item in question performs the functions that it was intended to perform. This is typically what the law requires, but most builders provide a warranty that extends beyond the minimal requirements of the law. This is done to improve client satisfaction, which should translate into client referrals.

Conflict Resolution

A final consideration is to address what remedies the parties will have if one of the parties fails to comply with the terms of the contract. Most contracts will typically call for some type of arbitration, and then, if the parties are unable to settle the dispute through arbitration, the only remaining remedy is the court. The builder should consult with an attorney to make certain that the company is adequately protected before signing any contract.

Transitioning to the Planning Phase

After the contract has been prepared, then it will be reviewed and modified by the attorney representing each party. Once all modifications have been agreed to and final copies have been executed, the time comes to transition to the planning phase, which is the second phase of the project lifecycle.

This typically involves giving the project manager the final approval and allowing him to gather individuals to work on developing the project plan. Once this has been done, the real work of planning the project is ready to begin.

REFERENCES

Kerzner, Harold. 2003. *Project Management,* 8th ed. Hoboken, NJ: John Wiley & Sons.

3

PLANNING THE
CONSTRUCTION PROJECT

The execution phase might be considered the most important phase of the project lifecycle. After all, the execution phase is where the actual work gets done and the home is actually built. This phase is the stage that everyone—the builder and the client—is typically in a rush to get to. Clients want to see progress; they want to see workers on the job site as quickly and often as possible. Construction managers also want to see workers on the job as often as possible. But before the work can begin a few questions need to be answered:

- In what order will the work be done?
- Who will be responsible for scheduling the workers?
- Who will perform the work?
- How will the subcontractors be selected?
- Will subcontractors provide their own materials, or will the general contractor provide the materials?
- Who is responsible for having the materials delivered?
- What is the budget for each item?
- What happens if someone goes over budget?
- Who inspects the work?
- What will be inspected?
- What happens if the inspected work does not meet the set standards?
- Who pays the bills?
- When are the bills paid?
- When can the clients inspect the job site?
- What happens if the client changes his mind?

These are only a few of the questions that are relevant to executing a construction project. The list could have been a few pages longer and still not exhausted all the questions that need to be asked. Someone has to have an answer for these questions, as well as the other questions that must be asked.

It is the job of the project manager to make certain that the right questions are asked and that the right answers are given. If a project manager does not do this, his job will be less like that of a contractor and more like that of a firefighter. Why is this so? Well, a firefighter puts out fires. The majority of a firefighter's job is responsive. Most of the time, firefighters are not called until the home is ablaze. They come to extinguish the flames. The project manager's job should be less like that of a firefighter and more like that of an estate planner. An estate planner is proactive by definition. This individual helps people plan for the eventualities that they will face. She looks at a person's life style, current financial needs, and future desires. From this information, she advises people how to get from where they are to where they want to be. This is a very proactive approach to living one's life—and it's a better model for managing construction projects. This chapter details how to move from being a reactive project manager to a proactive one.

OVERVIEW OF THE CHAPTER

This chapter contains the necessary knowledge and tools to develop a comprehensive project plan. First, an introduction to the nature of project planning is provided and directly applied to construction projects. This facilitates an understanding of the "big picture" of the planning phase. Second, an overview of those components that should be common to most project plans is provided:

1. Project scope and scope management
2. Construction schedule—work breakdown structure (WBS)
3. Cost control
4. Quality assurance
5. Human resource
6. Communication plan for team members and stakeholders
7. Risk management
8. Purchasing and contract administration
9. Project baselines

Following this overview is a discussion on how a project manager determines which components are necessary for different types of projects. This is important as, by definition, all projects are different. This is especially true in the construction industry. Therefore, not all projects will require the same type and size of project plan. Larger projects that are typically more complex will require a more detailed and comprehensive plan. Smaller projects will require less planning, as they are typically less complex. For instance, it takes more planning to construct a doctor's office than it takes to build a room addition on an existing structure. A detailed discussion of how to develop each part of the project plan follows, which will enable the reader to develop his or her own project plan.

NATURE OF THE PLANNING PHASE

The planning phase is the second phase of the project lifecycle. Figure 3.1 shows the five phases of the project lifecycle. The planning phase is different from the initiation phase in a number of ways, but one of the most important ways that it is different is that it is iterative. This means that the planning phase is continually revisited throughout the life of the project. The planning never truly ends until the project is completed and closed. Until that point, the planning phase will be continually revisited. Once the project has been initiated, the project manager moves to the planning phase, never to return to the initiation phase; once the project moves from the planning phase to the execution phase, the planning phase will be revisited as the project progresses and circumstances change. An example of this can be seen in Figure 3.2.

Figure 3.2 shows that once the project is being executed, the results are being monitored. In the controlling phase, the project manager attempts to determine whether the actual results match the planned results. If unplanned circumstances

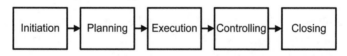

Figure 3.1 Project lifecycle

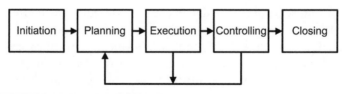

Figure 3.2 Nature of the project lifecycle

arise, the project manager must develop a plan to deal with those circumstances. A project that has been planned well from the beginning will have fewer unexpected circumstances arise, as compared to a project for which poor or no planning has been performed prior to project execution. Even when fewer unexpected events arise, they do still arise. This could be due to a number of factors: material shortages, changes in building codes, etc. Whatever the situation, the project manager will be forced to replan that portion of the project, which will then lead to a revised execution plan. The question, however, is what should the project manager plan during the planning phase? The next section of the chapter provides a brief overview of the components of the project plan, and then each component is discussed in detail.

OVERVIEW OF THE PROJECT PLAN COMPONENTS

Project Scope and Scope Management Plan

Recall that one of the steps in the initiation phase of the project is the development of a preliminary scope statement. The preliminary scope statement defines the goals and boundaries of the project. As a general rule, it contains the following items:

1. Why the project is being proposed—an internal or external project
2. What the project seeks to accomplish—building a house, a duplex, an office building
3. Who is responsible for the project—name the project manager

During the planning phase, the statement will be revised and expanded, if necessary. It is not always necessary to modify this phrase as the project unfolds if it accurately reflects the goals and the boundaries of the project.

Construction Schedule

The construction schedule is the document that will guide the entire project. In order to develop the construction schedule, the project manager must first develop a work breakdown structure (WBS) and the WBS dictionary, which seeks to name and define all tasks necessary to construct the structure. Developing the construction schedule also requires sequencing the tasks, developing time estimates, identifying the individuals or subcontractors responsible for each task, and developing a method to control the schedule.

Cost Control

The cost control plan accomplishes three primary tasks. First, it estimates the cost of each activity required to construct the building. The WBS is used as the guiding document. Second, it develops a budget for the entire project. Last, the cost control plan includes the information on how project costs will be monitored throughout the life of the project.

Quality Assurance

The quality assurance plan is composed of two primary parts. First, the quality assurance plan identifies the tools and techniques that will be helpful in ensuring the quality of the project. This includes identifying not only those methods that ensure that the structure is constructed according to a predetermined level of quality, but it also ensures that the project itself is managed well. Therefore, *quality* refers both to the quality of the output produced (for instance, the building) and to the quality of managing the project. Second, the quality assurance plan details a process for actually controlling the quality of the project. In order to do this, it identifies those individuals who will be responsible for using the selected tools and techniques to ensure the quality of the project.

Human Resource

The human resource plan is primarily concerned with how people will be hired, how they will be developed or trained, if necessary, and how they will be managed. This includes employees of the construction company and any subcontractors that might be hired to work on the project. Much of residential construction is out-sourced to subcontractors. Therefore, it is important to have a definitive approach to acquiring, developing, and managing subcontractor teams.

Communication

One of the most important aspects of any project is communication; this is especially true with construction projects. The communication plan provides a comprehensive guide to making certain that employees, subcontractors, vendors, customers, and other stakeholders receive communication in a timely and accessible manner.

Risk Management

The risk management plan is composed of two primary sections. In the first section, risks are identified and analyzed and responses are planned. In the second section, a plan is developed to monitor the various parts of the project in order to identify problems either before they arise or before they have done much harm to the project. It is important to note that although not all risks can be removed, many can be removed through appropriate planning.

Purchasing and Contract Administration

The purchasing and contract administration plan guides the purchasing and contracting of the project. The actual plan includes:

1. Purchase management
2. Subcontractor management
3. Contract management

Project Documentation and Project Baselines

This section of the project plan is primarily a reference section. First, it contains the documentation that guides the project, such as:

1. Project contract
2. Construction specifications—blueprints and building plans
3. Site survey
4. Building permit, insurance policy, septic tank approval and installation guidelines, sedimentation control plan, watershed control plan, etc.

Second, it contains the baselines pertinent to most construction projects, such as:

1. Cost
2. Schedule

These baselines provide a basis for comparison as the project progresses. By comparing actual results with planned results (that is, baselines), the project manager can determine whether the project is on schedule or within the cost guidelines. Failing to compare actual results with planned results can lead to unexpected delays or cost overruns.

CUSTOMIZING THE PROJECT PLAN

Because all construction projects are different, the project plan will need to be customized to match the needs of each project. This chapter identifies those portions of the project plan that should be common to most construction projects. However, the amount of time spent developing each part of the plan or the level of detail included in each specific section will depend on the project. For instance, if a construction company has a contract with a certain supplier to provide all building materials, the material acquisition plan will be rather simple. Or if a construction company always uses the same subcontractors, there will be no need to develop a plan to locate and interview potential subcontractors. Or if the project is a simple remodel or residential addition, portions of the plan may be unnecessary. Because of this, some time must be spent to determine what portions are necessary and how much information is to be included in each portion.

The first step is to determine which portions are necessary—or rather, which are not necessary. By default, every construction plan should contain the nine major sections previously mentioned. If the project manager or a stakeholder believes that one of the sections is unnecessary, the reasoning for such a position should be considered and decided upon. But by default, all nine sections should be included. The project manager and project team should also ask whether there are any special features to the project that makes another section necessary.

By including all nine sections by default, the project manager can assure that a comprehensive approach to planning construction projects is being taken. Going through each of the nine sections helps to ensure that no portion of the project is being disregarded or overlooked. This way, the project manager is purposefully considering the project in its entirety. This is a difficult task because it requires the type of work that many project managers do not enjoy. They enjoy being in the field, working on the job site, but by spending the time to develop this project plan, they can help ensure project success.

The second benefit of including all nine sections is that the project manager can work to eliminate the problems that often arise from assumptions. The project plan puts the assumptions of the project into writing. It provides an equal amount of knowledge and understanding to the entire project team, which will help eliminate those mistakes that can often arise when people attempt to manage a project from their head or on the back of a napkin. This second point also introduces a topic that needs to be addressed—information hoarders.

This may sound like a strange phrase, but it is something that just about everyone will experience if working in an environment with enough people. An information hoarder is someone who attempts to be the single repository of all information. She only releases information on a "need-to-know" basis and, to her, most people do not need to know. This type of person is always vying to be the go-to person; she wants to meet with the subs, the customer, or any other stakeholder she can. This type of person can be devastating to the success of a project. A comprehensive project plan that provides information about the project is an excellent means of combating this type of person by allowing information to be accessible.

This does not mean that all information should be available to all people. For instance, most companies have a vested interest in keeping their profit margins internal. Portions of the project plan may be restricted, but, in general, the more information that is available, the better the chance that no one will make a wrong assumption or guess that could create a costly mistake.

DEVELOPING THE PROJECT PLAN

The project has been researched, the project manager has been assigned, the construction documents have been compiled (blueprints, material specifications, etc.), and the time has come to actually develop the project plan. Developing the project plan takes time, especially for those who have never attempted to develop such a comprehensive plan before. There is a learning curve. The first time will be the most difficult and time consuming. But time spent at this stage will save a great amount of time later on when the building is actually being constructed. Therefore, take the time to plan the project properly. The next time it will be easier.

Each section of the project plan is covered in some detail, although not everything that could be said is discussed. Entire books have been written on topics that are covered with a few paragraphs in this book. But this guide offers enough information for project managers to develop working project plans that can be used for their construction projects.

Project Scope and Scope Management

The scope of the project refers to what is and what is not included in the project. During the initiation phase of the project, a preliminary scope statement was crafted, expressing an overview of the project. This statement—along with the construction contract, blueprints, building specifications, and other similar documents—acts as a guide to developing this portion of the project plan.

This portion of the project plan will focus on accomplishing the following tasks:

1. Finalizing the scope statement
2. Developing the work breakdown structure (WBS) and WBS dictionary
3. Developing the scope management plan

Finalizing the Scope Statement

The scope statement is finalized by assuring that the preliminary statement accurately reflects the goals of the project. This is done by comparing the preliminary scope statement to the construction contract and related documents. For instance, if the client was initially requesting a 3500 square foot home, but during contract negotiation modified the floor plan to 3000 square feet, then the preliminary scope statement would need to be updated to reflect the change.

An important purpose of the scope statement is to remove ambiguity that may be present elsewhere. Therefore, the scope statement is not only about saying what is included, but equally about saying what is not included. For instance, if the project manager's construction company does not provide landscaping services, the scope statement should state that those services are not included. Many clients make all sorts of assumptions about what will be included (the mail box, window blinds, etc.). The scope statement attempts to spell out for them what services and products the construction company is actually providing. Many times, this can be accomplished through a standard list of features and services, which can be attached to the project plan and simply noted in the scope statement. Once the scope statement has been finalized, the project manager is then ready to develop the work breakdown structure (WBS) and WBS dictionary.

Developing the Work Breakdown Structure and WBS Dictionary

The WBS is a chart that shows the work necessary to accomplish the project. It is a chart that is most easily understood by seeing an example, as shown in Figure 3.3.

One purpose of this chart is to show how all the work necessary to accomplish the project relates to the project as a whole. If someone were to see only the WBS of the project, he would gain a very good idea of what the project entails. Obviously, there would be numerous details still necessary to actually execute the project, but the WBS provides an excellent overview of the project. Therefore, one of the purposes of the WBS is to show how the project is internally related.

A second purpose of the WBS is to help facilitate the planning process, as it is a very valuable planning tool. The WBS and the WBS dictionary, which will be discussed shortly, provide a wealth of information for those planning the project. It tells those scheduling the project what activities they need to schedule. It tells those performing cost estimates what costs need to be estimated. The WBS dictionary is a closely related document, which provides details about each activity on the actual WBS.

Although developing the first WBS for each type of project for a company takes some time, once a project manager creates a WBS for a certain type of project, she can simply modify the previously used WBS to match each new project's specific needs. WBSs developed for future projects should take considerably less time.

The steps to develop a WBS are:

1. Gather necessary documents (building plans, blueprints, material and feature specifications, building contract, etc.)
2. Gather necessary people (project team, general contractor, subject matter experts, etc.)

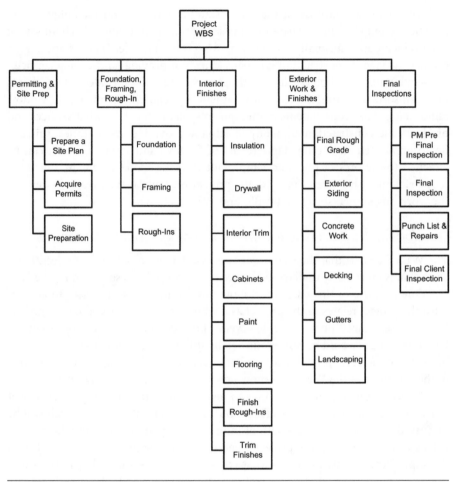

Figure 3.3 Sample WBS 01

3. Create a high-level structure of the project
4. Add low-level details to high-level structure
5. Create WBS dictionary

The first step is to gather those documents that will be helpful to creating the WBS, for instance, the finalized scope statement, the construction contract, the building plans, any material or feature specifications lists, and any other documents that provide input into what the project will actually produce. Once the project manager has gathered the proper documentation, she brings together those individuals who will be responsible for developing the WBS. If this is a large construction project, the team comprises the number of individuals that the project manager believes to be necessary to accomplish the task. For a smaller project, the project manager may be the only person to work on the WBS. Regardless of the size of the team, it is important that those who are involved are given copies of the information that was gathered in the first step so that they can familiarize themselves with any and all information that relates to the project. After the plans and specifications have been gathered and the team has familiarized themselves with the information, it is time to develop a high-level WBS structure. The WBS should be only as detailed as is deemed helpful for planning the project. The goal is to determine how the work should be arranged. Consider the following example.

The project team has been brought together to develop a WBS for a four-bedroom, two-bath, 3000 square foot home. The team has copies of the construction contract, blueprints, building specifications, and other related documents, which provide an overview and perspective on the project. Let's assume that the building contractor will be responsible for the entire construction project except the final landscaping. In order for the project team to develop an accurate WBS, they would need to know this information. If they include something that is not supposed to be included, it could create major problems later on in the project.

After the team has familiarized themselves with the project documents, they are ready to complete this high-level WBS structure. Some WBS structures are more simple than others. Figure 3.4 shows a sample WBS, which has only one level of detail. This high-level structure would correspond to the Level One in Figure 3.4. Below Level One, additional levels would be inserted, which would increase the depth of the WBS. There is no rule about how wide or how deep the WBS should be; the project manager and team must use their best judgment. In determining the structure of the WBS, the project team is attempting to discover the point at which the WBS becomes overly cumbersome and becomes a hindrance. The WBS that I employ has four levels of detail, but for others, four levels might be either too much or too little.

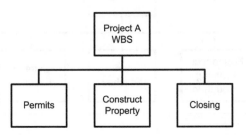

Figure 3.4 Sample WBS 02

Others will create a more detailed WBS structure in which the first level has more indicative detail, such as is shown in Figure 3.5. This second example is much more detailed than the first, but it is also a little broader, which may or may not be helpful for a specific project. The project manager and team must consider each specific project and adapt the WBS to meet the needs of the project. The flexibility of this tool is one of its greatest assets in the planning process.

After the high-level structure of the WBS has been developed, the project team is ready to add the lower-level details to the skeletal structure of the WBS. This is done by thinking through the various stages of the WBS and inserting those tasks that must be accomplished in order for the project to be completed. As an example, take a look at Figure 3.6. Figure 3.6 shows the lower-level structure associated with the permitting and site preparation phase identified in the WBS structure of Figure 3.5.

Notice in Figure 3.6 that the WBS provides no reference to the time that it will take to perform the tasks, nor to the costs that are associated with the tasks, nor does it state who will be responsible for performing the tasks. This is because the WBS is merely a planning tool to guide many of the following steps.

For clarity in following all the steps in a WBS, the WBS is coded. Coding is the means by which the project manager assigns a number or value to each element of the WBS to facilitate clear communication about the WBS. For instance, if the project manager wanted to reference a particular element of the portion of the WBS presented in Figure 3.6, then she would have to describe that element in relation to the other elements in a rather cumbersome manner. For instance, if the element to be discussed was *walk lot*, the project manager would have to say something like the following: "Element entitled 'walk lot,' which is the second lot

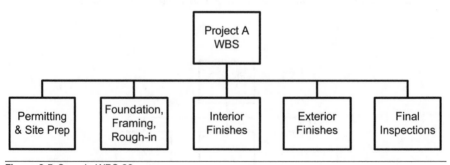

Figure 3.5 Sample WBS 03

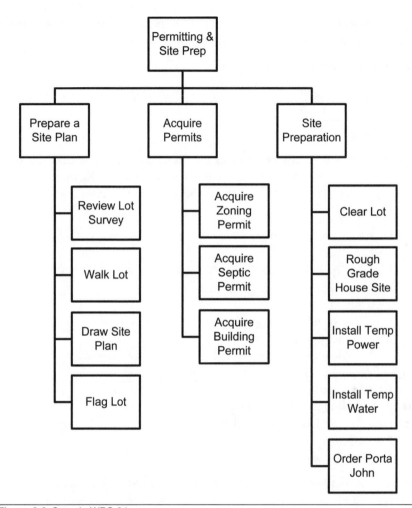

Figure 3.6 Sample WBS 04

element below the site plan element, which is the element to the far left of the permitting and site prep element." This is a rather cumbersome method of identifying the various elements of the WBS.

Best practices prescribe that a coding system should be employed in developing the WBS, which allows the WBS dictionary and any other project documents to easily reference the various elements of the WBS with minimal confusion.

Using Figure 3.6 as an example, one can easily see how a coding system can be applied to a WBS. In this example, the pinnacle of the chart is assigned as Element 1 or simply 1. The tasks below it are assigned the following codes:

1.1—Permitting and site prep
1.2—Foundation, framing, rough-in
1.3—Interior finishes
1.4—Exterior finishes
1.5—Final inspections

The period between the numbers informs the reader that a lower level of the WBS is being referenced. The labeling for subsequent levels is similar. Again referring to Figure 3.6, the elements in that chart should be assigned the following codes:

1.1.1—Prepare site plan
    1.1.1.1—Review lot survey
    1.1.1.2—Walk lot
    1.1.1.3—Draw detailed site plan
    1.1.1.4—Flag lot
1.1.2—Acquire permits
    1.1.2.1—Acquire zoning permit
    1.1.2.2—Acquire septic permit
    1.1.2.3—Acquire building permit
1.1.3—Site preparation
    1.1.3.1—Clear lot
    1.1.3.2—Rough grade house site
    1.1.3.3—Install temporary power
    1.1.3.4—Install temporary water
    1.1.3.5—Order porta john

The above example illustrates how a coding system should be developed. Someone might choose to use a slightly different system, which is not necessarily problematic. The important point is that a system is employed to ensure that reference can be made to the various elements of the WBS with minimal confusion. Best practices also state that the codes should be inserted into the WBS chart beside the element it identifies.

After the WBS has been developed, both the high-level and the lower-level detail, and coded, the project manager and team is ready to prepare the WBS dictionary. The WBS dictionary performs three critical functions:

1. The WBS dictionary defines each element of the WBS
2. The WBS dictionary describes the resources necessary to perform the task
3. The WBS dictionary prescribes the processes to follow when carrying out the work described

In the WBS dictionary, the code for each element is listed and the associated element is described. The detail of the description will depend on what the project team believes to be necessary. In general, however, the description should include a basic summary of the process. As mentioned earlier, the WBS and the WBS dictionary are used extensively during the planning phase to develop the project plan. Think of them as the grand repository of raw project information. Along with a description of each element, a description of the resources necessary to accomplish the task should be included, as it will greatly help later planning. For instance, in acquiring the septic permit for a construction project, it is often necessary to clear a portion of the lot and dig perk holes for the environmental health inspectors (soil scientists); if this will be necessary, the description should state it, and the dictionary should also include those resources necessary to clear the portion of the lot and dig the holes that will be required. This allows those who are referencing the WBS and WBS dictionary for planning purposes to make certain that the necessary resources are both allocated and available when required.

Finally, the WBS dictionary prescribes the processes that should be followed when carrying out the elements found in the WBS. This portion is sort of a special notes or best practices section. Often, there are numerous ways to accomplish the same task. Different contractors and subcontractors have various ways of performing the same work. If specific guidelines need to be followed, the WBS dictionary should say so; otherwise, the worker will perform the work as he sees fit, which may or may not give rise to a future issue.

To make certain that these three elements are included in the WBS dictionary entries, all entries should include the following information:

- WBS code
- Element description
- Required resources
- Instructions/special notes

An entry might appear as follows:

WBS Code: 1.1.1.3—Draw detailed site plan

Element Description: The detailed site plan should include the following items:

- Location of home
- Location of driveway
- Location of septic field
- Location of power, water, sewer, cable, and other utility lines
- Building setback lines
- Any right-of-ways or easements crossing the property
- General location of landscaping
- Erosion control measures (silt fences, etc)
- Elevation of key points of lot (front, rear, house site, septic field, etc.)
- Distances from lot lines to home, driveway, septic tank and field, and any other improvements to property

Required Resources: House plans, lot survey

Instructions/Special Notes: Plan should be drawn to a 1/40 scale; reviewed by project manager for final approval.

By reviewing the above entry, a project team member should be able to either create the site plan or locate someone who has the particular skills necessary to do so. She should also be able to estimate the amount of time it would take to create such a plan. The purpose of creating descriptions such as this for each element on the WBS is to gather the appropriate information to make informed decisions so as to avoid costly mistakes when the actual project is being executed.

Developing the Scope Management Plan

Under project scope and the scope management plan, the scope statement, WBS, and WBS dictionary are primarily concerned with project scope. The second part to consider is the scope management plan that instructs the project team in how major change orders should be handled: for instance, if the customer or another key stakeholder wants to make changes to the scope, how those changes will be planned, costed, and implemented.

The scope management plan is not concerned with simple change orders, but with major changes in the project. For instance, let's assume that after the contract has been approved and planning is underway, the client requests that the builder construct a different style of home than the one quoted and agreed upon. How will such a request be handled? Who can approve such changes? In most construction companies, only a very senior member of the company can approve such a change, but this should be stated; the project plan should identify who can approve such changes, and what the process is to make such a change. Many times a company policy will address such issues, which can be included as part of the project plan. If there is not a company policy, then a policy will need to be developed for each particular project.

The definition of a major change is relative to the project being undertaken. If someone were constructing a $100,000 home, and the client wanted to add a $30,000 outdoor pool to the project, such an addition would be a major change. If, however, the buyer wanted to upgrade the front door, which would cost an additional $300, this should probably not be classified as a major change. Later on, these lesser types of change order requests will be addressed.

Construction Schedule

When those who have never developed a construction schedule look at the final product, they can rarely appreciate all the time and effort that has gone into it. Developing the construction schedule is one of the more time-consuming portions of the project plan. It requires much more than simply stating that this must be completed before that or that this will take three days to do and that four days to do. It involves carefully considering each task of the project and creating a road map that will lead to project success. The steps to create a project schedule are:

1. Define tasks
2. Sequence tasks
3. Estimate required task resources
4. Estimate task duration
5. Develop schedule
6. Manage the schedule

Define Tasks

The first step is to define the tasks or activities required to construct the building. One might wonder if this has been done already in the WBS and WBS dictionary. Yes and no; it depends. One factor to consider is how detailed the WBS and WBS dictionary actually are. Those who have developed less detailed WBSs will find it necessary to define some additional tasks at this point than those who have more developed and detailed WBSs.

Two additional points to be considered are task complexity and involvement. Tasks vary in their complexity—some are simpler and some are complex. Consider creating a concrete foundation footing for a standard 1400 square foot rancher. If the soil quality is good and the ground is mostly level, in most cases, this will be a rather simple task, which can be completed by a skilled crew in no more than a day or two. However, the same task might take much more time depending on the size of the home, the quality of the soil, the elevation of the land, and other factors. After an initial inspection of the site, the project manager might determine that the footing should be a fairly straightforward part of the project. In that case, the WBS and the WBS dictionary might provide ample instructions, as they will specify installation of the footing in accordance with the design of the plan and local building codes. However, if the project manager believes that the footing will require special considerations, either because of design or environmental issues (rock formations, unstable soils, etc.), the task will need additional definition, as it will require special attention due to its complexity.

The second consideration is involvement, which is closely related to complexity. The question to be answered at this point is, "What is the project team's direct involvement with the task in question?" Construction companies vary in the amount of construction work they actually do. Many companies have direct employees who perform almost all of the construction tasks: footings, framing, roofing, plumbing, electrical, and so on. Other companies subcontract out almost all of the construction tasks. The companies with a large number of employees who actually perform the work of the construction project will inevitably plan in greater detail than those who use more subcontractors. Therefore, if the construction company itself were planning to install the more complex footing considered above, it would need to develop what might be called a footing miniproject in order to make certain that all of the work was performed as required. It would not be enough for the project manager to simply schedule the footing; instead the project manager would want to outline those tasks that would allow the construction crew to know what it is going to take to actually construct the footing.

If, however, a construction company that subcontracts out its footings were faced with a complex footing, they would simply ask the subcontractor to estimate the time required and the cost of constructing the footing according to acceptable standards. In this instance, the project manager then plugs those numbers into the project plan. In this case, there is no additional planning beyond simply inserting the footing step into the project schedule with the appropriate parameters. The project team would have a lower level of involvement than in the previous scenarios.

Figure 3.7, which can be used as a risk management planning tool, provides a visual means of evaluating tasks. Using it as a guide, the project manager can

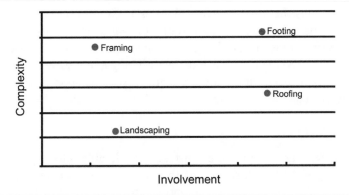

Figure 3.7 Task complexity and involvement

determine which tasks will require definition beyond the WBS and the WBS dictionary. Consider four tasks of varying degrees of complexity and involvement:

- Footing—high complexity, high involvement—quadrant one (upper right)
- Framing—high complexity, low involvement—quadrant four (upper left)
- Roofing—low complexity, high involvement—quadrant two (lower right)
- Landscaping—low complexity, low involvement—quadrant three (lower left)

Because of a number of environmental and structural factors, the footing for the project is considered to be a complex task. Also, the construction company does not outsource footings, but uses internal employees. Therefore, the task would be placed in the high involvement, high complexity quadrant. A task that is plotted in this quadrant will most likely require additional definition beyond the WBS and the WBS dictionary. Instead of having a task labeled "install footing" on the project schedule, there will most likely be a task labeled "footing," which has a number of subordinate tasks, such as: (1) order engineer's report; (2) review engineer's report with construction supervisor, building inspector, and engineer; (3) dig footing per engineer's report, and so on. The point being that because of the amount of direct involvement by the construction company's crews and because of the complexity and importance of the footing for the structural integrity of the home, it is vital that adequate planning be performed to ensure that the footing is handled correctly.

The second example is framing. In this example, the construction company outsources framing to a subcontractor. Therefore, framing involves a high degree of complexity, but low involvement on the part of the construction company. The input for the construction schedule will come from the framing subcontractor, who will provide a timetable for performing the work. The project manager will most likely want to insert into the schedule inspections while the work is being performed, given the high level of complexity. Using a tool such as this chart will cue the project manager and planning team to pay the appropriate amount of attention to each task.

Sequence Tasks

The second step in developing the construction schedule is to sequence tasks in the order they will be performed. For the most part, sequencing tasks is fairly straightforward, but there are more variations than one might imagine. Early on, the schedule will be fairly straightforward, but once the home gets framed and under-roof, in some cases, there will be essentially two simultaneous schedules running concurrently for the internal and external portions of the home. The project manager needs to consider two things: how to develop the schedule and how to record or visually represent the schedule. In regard to the project, the former is more important, but knowing the latter will help with developing the former. Therefore, first attention will be turned to how to put the schedule together or visually represent the flow of work.

The following suggests a few options for putting the schedule together:

- Pencil and paper
- Nonscheduling software—spreadsheet or word processing
- General scheduling software—e.g., Microsoft Project©
- Construction-specific scheduling software

The first option is the old-fashioned way; making a list of the tasks in numerical order with pen and paper. The problem with this method is that a list of the tasks in numerical order, for most projects, fails to capture the true flow of the project. For instance, it is difficult to show how two tasks can be occurring concurrently by simply listing the tasks from the beginning to end. Instead of listing the tasks by number, one could attempt to create a flowchart. This is preferred to simply listing the tasks, as it will provide an accurate picture of the relationships that the tasks have with one another. This is a time-consuming method, difficult to modify as changes are made to the project. There are a number of methods that have been presented as ways to show the relationships between the various tasks, such as the precedence diagramming method, the arrow diagramming method, and a host of other schedule network types. Explaining each of these methods is beyond the scope of this book, but most project management or system management textbooks contain explanations of various methods.

A second option is to use nonscheduling software. Most people have access to word processing or spreadsheet programs. Using these tools is definitely a step up from the first method, but it has its own challenges as well. Primarily, creating a flowchart is possible, but it is time consuming. Once it is built, it is easier to modify, but major changes can result in much wasted time.

The third option is to use general scheduling software, which is to say non-construction-specific scheduling software, such as Microsoft Project. I use Microsoft Project. The benefits of using actual scheduling software far outweigh the cost. Microsoft Project is just one of many scheduling programs. In Microsoft Project, the tasks can be easily entered and sequenced, and the software will automatically create a sequence schedule, which can be printed or E-mailed for review. Microsoft Project is a very powerful tool that contains a number of powerful features that can take a long time to master. However, even a novice with a simple guide can reap great rewards.

The last option is to use construction-specific scheduling software. This type of software works similarly to Microsoft Project, but is tailored with features for the construction industry. Most packages include much more than scheduling software, offering accounts payable, time management, and a host of other features. Carefully consider any purchase and make certain that a demo is available.

Once the project manager has determined how to record the schedule, it is time to develop the sequence in which the work will be performed. This should be a straightforward task, as it is fairly obvious how much of the work should be scheduled. The WBS, WBS dictionary, and task definitions completed in the first step are the primary documents to reference. Other than that, the project manager should rely on her judgment or the expert opinion of others in developing the schedule. Figure 3.8 is a sample flowchart showing a simplified portion of a project for constructing a single-family dwelling.

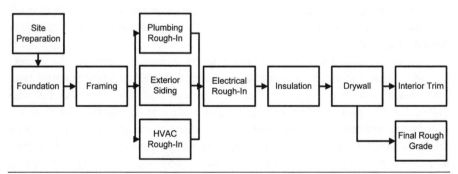

Figure 3.8 Sample schedule flowchart

In the example, each task is listed along with the task's predecessor. Most of these will be firmly fixed, as each task builds on others. There is some flexibility; how much will depend on local requirements or builder preference. For instance, an electrical company might require that vinyl siding, if used, must be installed prior to doing the rough-in wiring, which ensures that a nail does not split a wire. This is not a must, but it is the preference of the subcontractor. The company for which I work will not pour a concrete driveway until the drywall has been delivered to the site, which is the last of the heavier duty deliveries to be made to the job site. These are not musts, but they are examples of the types of issues that must be considered while scheduling the required work.

The best way to discover particular requirements or preferences is to ask the company's employees and subcontractors. The project manager should ask them at what point of construction they want to come to the job site and if they have any unique requests. If they do, the project manager should learn why. If it is a better practice than was being planned, he should implement it. Many things look good on paper, but will not work on the job site. Using the expert advice of the workers on the ground results in a schedule that is developed in the office and will work in the field.

Estimate Required Task Resources

The next step is to estimate the resources necessary to accomplish the work. What will it take to do the job? For the tasks being outsourced to subcontractors, this will not be necessary, as the subcontractor will be responsible for gathering her own resources. But if it is a new subcontractor, then it is important during the interview process to ensure that the subcontractor has adequate resources to complete the work required. For instance, if a subcontractor is being considered for performing the initial site preparation but does not have equipment large enough to perform the job, this is likely a deal breaker. In general, the project manager will not be required to estimate the resources for portions of the project that are subcontracted out; he must only ensure that the subcontractor is able to perform the job as agreed.

For those tasks the construction company is doing internally, the project manager must accurately estimate the resources required to perform the task. The information for doing this will come primarily from the WBS and the WBS dictionary. The project manager will need to make certain that the company has adequate funds, equipment, and workers to perform the work. The best way to do this is to have a list of each task the company will be performing, a list of equipment the company has access to, and a list of workers who can be assigned. Using the task list as the guide, the project manager goes through and assigns equipment and workers to each of the tasks.

In order to do this, the project manager must have knowledge of what equipment it will take to perform the task and knowledge of what skills the workers will need to possess. If she lacks that information, she must get it in order to accurately estimate resources. If she finds that the company lacks either a particularly skilled worker or a piece of equipment for the job, this can be remedied in one of a couple of ways. First, the project manager can request that a new employee be hired. If the need at hand is going to be a recurring one, then this might be an option. Second, the project manager could seek out another subcontractor to perform the work. If a piece of equipment is lacking, then the project manager can request that it be purchased or rented, or, again, a subcontractor can be employed to perform the work. After resources have been assigned to each task, the project manager is ready to estimate task duration.

Estimate Task Duration

Estimating task duration is a critical part of developing the project schedule. The project manager now has a rather large amount of information to consider before assigning a duration estimate to each task:

- WBS
- WBS dictionary
- Task definitions
- Task sequence
- Task resource estimates

Using this information as well as expert opinion (his own and those around him) and historical data (past construction projects), the project manager is ready to estimate how long each task will take.

There are a couple of different approaches that can be taken when estimating how much time to assign to each task. Some figure a best case and a worst case scenario and go with the average. Some estimate a best scenario and double it. Some companies add in a standard amount of safety time, in case something arises. There are also a number of more complex methods for developing duration estimates. Some use the critical path method; others argue for the critical chain method, based on Goldratt's theory of constraints. The more complex the project, the more complex the method typically is.

All of the above methods, to some degree or another, will produce a schedule that looks great on paper, but what happens when construction begins? What happens to the schedule when flooding in the Midwest delays shipments of brick, or when a subcontractor who was hired for the job declares bankruptcy, or when an employee fails a drug test, or when it rains for a week, or when one hits a water table or underground spring when digging for the footing or the septic tank? A construction project is not performed in a controlled environment; it is open to human complications as well as environmental complications. Someone who develops a schedule without attempting to take into account the complexities and factors of a construction project will probably be frustrated more than anything. All this is complicated even more if a builder is constructing multiple projects. Consider, for instance, the company for which I work. At any given time, the company has 35 to 40 homes under construction with anywhere from 30 to 60 percent sold, and the remainder being spec homes. How can a workable schedule be developed for each individual project that takes all construction projects into account? A schedule that offers the flexibility required for the construction project must be developed. In this section of the book, developing the task duration estimates for a single project is considered.

A common method for developing the task duration estimates is to assign a realistic assessment of the time each task will require given optimal working conditions. For instance, for most projects I engage in, a footing should take no more than four to five days for an average single-family residential home given normal working conditions. Therefore, the estimate would be five days. A project manager would then go through each task; consider the complexity of the work, the resources required, the ability of the workers, and the input from the workers on what they believe it will take; and then assign a reasonable duration to the task.

The project manager will also need to get time estimates from the subcontractors. These must be critically assessed to ensure that they are accurate. Some subcontractors will underestimate in hopes of getting the bid. Others will overestimate, if they believe that they possess a stronger bargaining position. In considering a subcontractor's estimate, the project manager relies heavily on past experience with the subcontractor and reviews provided by the subcontractor's other customers.

The individual estimates created for each task and those gathered from the subcontractors are most likely reasonable individual estimates considered on their own. However, if one were to consider the schedule of the project as merely the sum of its tasks, then a very unworkable schedule will be created.

One method that can be used to estimate task duration, is called three-point estimating. A three-point estimate relies heavily on the statistical method called the normal distribution. In order to perform this method, the project manager will prepare three time estimates for each task, which are labeled as follows: best case estimate, most likely estimate, and worst case estimate. These numbers are then inserted into a statistical formula which produces an estimate.

Let's say that a project manager is developing a task duration estimate for the framing of a construction project. If the three-point estimate were to be used, then three estimates would need to be given as:

- Best case estimate (a), 5 days
- Most likely estimate (m), 7 days
- Worst case estimate (b), 10 days

A three-point estimate is obtained by calculating E, which represents the value of the estimate, using the formula:

$$E = \frac{(a + 4m + b)}{6}$$

Using this formula and the estimates above, in this example $E = 7.167$. According to this method, one should round up to eight days. The value of this method is that it takes into account the best and worst case scenarios, but this does not mean that the task *will definitely* take just more than seven days. It means that based on this statistical model, it will *most likely* take just more than seven days.

A companion formula is used to calculate the standard deviation, which does not alter the estimate given above, but it provides an estimate of the certainty that the task will take the time yielded by the calculation. The standard deviation (SD) is calculated as:

$$SD = \frac{(b - a)}{6}$$

There is much more that could be said about this method; a more complete discussion can be found in almost any general project management textbook.

Another method that can be used is based on the critical chain methodology briefly mentioned earlier. In the next section, the reader will find a detailed discussion about how this method both creates schedules and estimates task durations.

Create Schedule

At this point, the project manager is ready to create a construction schedule. At hand, the project manager will have an abundance of information that can be used for creating the project schedule. The project contract will specify what the anticipated start date of the construction is to be. The WBS and the WBS dictionary provide information on what work is to be performed. The tasks have been further defined, the resources to perform each task have been estimated, the time needed to complete each task has been estimated, and the tasks have been sequenced. Now it is time to put all this information together into a workable and flexible schedule.

Scheduling is an iterative process because a variety of factors begin to affect the project once construction begins: weather, material shortages, subcontractor delays, and so on. If the schedule is developed properly, then it will anticipate that certain situations will arise and will need to be managed. Amendments will most likely be needed, but they should be seen as slight adjustments. The schedule should not need to be drastically altered if it has been appropriately created.

This first schedule will act as the schedule baseline for the project. As the project progresses, the schedule will require regular updates, but the original schedule is preserved. The actual results will then be compared to the original schedule, the baseline, which will allow the project manager to see how well the project was planned. This will provide feedback for developing better schedules in the future.

There are a couple of major schedule types that one can employ in creating a schedule: the critical path method and the critical chain method. Although these methods may seem similar at first glance, they are actually quite different. Both methods are presented as options, and the strengths and weaknesses of each method are addressed.

Critical path method The critical path method (CPM) is a relatively straight-forward project management tool, which seeks to identify the longest sequence of tasks to accomplish a project. This sequence is termed the *critical path*. For instance, consider the flowchart depicted in Figure 3.9.

Figure 3.9 represents a project that requires six tasks in order to be accomplished. Those tasks which are shaded gray represent the critical path, as they are the sequence of events with the longest duration. The critical path is 42 days in duration; whereas the alternative path—being Task A → Task B → Task E → Task F—is 38 days in duration. The alternative path is, therefore, not the critical path, because it is shorter. One project can only have one critical path at any given time.

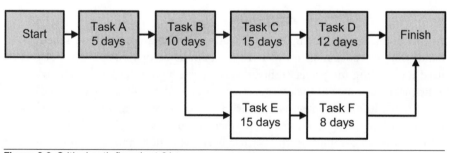

Figure 3.9 Critical path flowchart 01

A project's critical path can change. For any number of reasons (material shortages, labor shortages, weather delays, or increased worker productivity), the original critical path can shift from one sequence of activities to another, as the critical path is the longest sequence of events at a given time. For instance, if due to weather delays and material shortages, the duration of Task E were extended to 30 days, the new critical path would appear as shown in Figure 3.10.

Now Figure 3.10 shows the critical path as being 53 days, compared to the original critical path, which was only 42 days. Using this method, the project manager is continually making certain that the correct critical path is known and that the management of the project is based on the current critical path.

Note that the critical path method *does not take resources into consideration*. For instance, Figures 3.9 and 3.10 show Task C and Task E as starting at basically the same time. What the chart does not show is that the same person will be performing those two tasks. Although the tasks are not immediately dependent on one another, they are dependent in the fact that only one person on the project team is capable of performing the tasks. Therefore, instead of the project taking 53 days, as projected by the critical path method, it will actually take 72 days, because Task C and Task E cannot be performed simultaneously. Before this project has even begun, it is already 19 days behind schedule. This is shown in Figure 3.11.

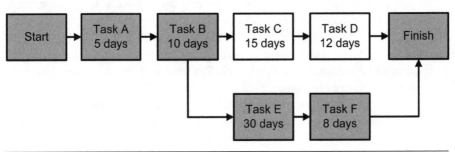

Figure 3.10 Critical path flowchart 02

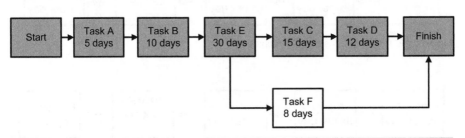

Figure 3.11 Critical path flowchart 03

Critical chain method Whenever a project schedule takes resource constraints into account, it is called *resource leveling*, as it levels out any resource conflicts. After the resources have been leveled, the project manager has a schedule that more closely resembles the project's critical chain. The critical chain of a project is the longest sequence of activities in a network after specific consideration has been given to the availability of project resources. Figure 3.11 shows the project flowchart as adjusted to show the critical chain of the project.

The critical chain method of developing a project's schedule is based on Eliyahu Goldratt's theory of constraints (TOC). The TOC is a method of management that seeks to locate and remove constraints that hamper efficiency and productivity. It was originally developed for application in a manufacturing environment, but numerous applications have been found in the field of project management, as well as others.

As an example of TOC, consider a widget manufacturing process. Five processes are required to produce the widgets, and each process varies in the time required to complete. Figure 3.12 shows the process in a flowchart format. Process 3 is the most time intensive, as it can only be performed six times per day. If this production schedule were to hold true, only six widgets could be produced per day regardless of how efficiently Processes 1, 2, 4, and 5 were performed. Therefore, Process 3 is known as the bottleneck of this production schedule.

If the production manager attempts to run each process at full capacity, a backlog of two widgets per day would be created unless the production capacity of Process 3 were increased by some means. If the production manager were to improve Process 3 so that it could be performed nine times per day instead of six times, it would no longer be the bottleneck. Process 1 and Process 5 would become the bottleneck, as seen in Figure 3.13.

In the revised situation, only eight widgets per day could be produced until the production capacity of Processes 1 and 5 were altered. If their production

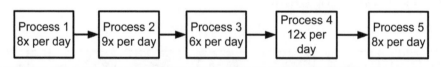

Figure 3.12 Theory of constraints flowchart 01

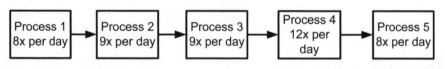

Figure 3.13 Theory of constraints flowchart 02

capacity could be increased, it would create a new bottleneck, which would then need to be analyzed and modified. This is a form of continuous improvement.

The application of such a theory to the production of simple widgets that require only five processes to manufacture is easy to see, but the application to project management may be less obvious.

Critical chain project management (CCPM) is the result of the TOC being applied to the field of project management. Most practitioners of project management are familiar with the critical path method (CPM) of project management, but less familiar with CCPM. This next section will provide the briefest outline of the CCPM.

CCPM is different from the CPM at a couple of fundamental levels. First, when developing the project schedule, CCPM takes resource constraints into account from the beginning. In the CPM, the schedule is developed and resource leveling is an added step, not really part of the method. If this step is ignored, the schedule developed will be unrealistic. CCPM, however, levels resources as part of developing the project schedule. Leach, an expert in CCPM, states that "the critical chain is the longest set of dependent activities, with explicit consideration of resource availability, to achieve a project goal" (Leach 2005, 243). In this way, the project manager is assured to have a schedule that does not have conflicts between the various workers and subcontractors.

Another difference between the methods is that through the use of resource and cost buffers, CCPM offers a new method to determine project efficiencies, which is not available with the CPM. Finally, CCPM uses a modified method to estimate task duration, based on basic human behavior, which CPM ignores. These behaviors include Parkinson's law and the student syndrome.

Traditionally a project manager uses the WBS to develop a task duration estimate for each task required to complete the project. When this is done, the estimator will consider weather, material delays, schedule delays, possible work complications, and other such things that lead to the estimate that containing a considerable amount of padding. If someone could actually work solely on one task, it might take three days to complete the task. But the person expects to be interrupted or delayed, so a few extra days are added to the estimate. Then when the estimate is turned in to the project manager, a certain percentage buffer might be added on top of everything else as a safety net. What is produced is a project schedule that can contain a few extra levels of padding. The goal of such padding is to protect the project completion date from uncertainty or apparent delay. But when each task is padded with this extra time, it can create a bloated schedule.

When this extra padding is considered in light of basic human behavior, one can easily see how this produces a bloated schedule. A project manager might think that if her projects are padded a bit, it is a good thing because she usually finishes the projects with no time to spare. This is most likely true, but it should be considered in light of a few behavioral observations.

First, Parkinson's law, which is named after British historian and author C. Northcote Parkinson, states that work expands to fill the time for its completion. Simply stated, people will typically take all the time allowed to complete their tasks. Therefore, if the time required to complete a task is two days, but the worker has four days to complete the task, then one should assume that the four days will be taken.

A second behavioral observation to be considered is student syndrome. Student syndrome refers to the phenomenon that people will wait to apply themselves to a task only on the brink of the deadline, which I can attest to as I am writing this chapter. Therefore, people knowing that they have a certain amount of padding built into their estimate will often wait until the last possible minute to complete the task.

The effect of these factors is compounded when one considers the poor job that most people do when attempting to complete multiple tasks at once. Typically, if employees or subcontractors have three tasks to accomplish, and each task takes about three days, and they work on each task an even amount of time per day, then the three tasks will take a total of nine days, as the workers are only applying themselves to a particular task one-third of each day. Multitasking is clearly unproductive, but most people would say it is necessary. And it may very well be necessary because workers are answering to multiple bosses who have developed poor schedules that do not take each person's other responsibilities into account. If the multitasking could be eliminated, then one task would be completed only at the end of nine days, but the other two would be completed in three days and six days, respectively.

Now that some of the reasons that estimates are padded have been discussed, it is time to begin adjusting task estimates according to the principles of CCPM. The first step is to adjust all the task estimates and insert project buffers. In order for the reader to see the differences between a CPM schedule and a CCPM schedule, it might be helpful to review what a CPM project schedule might look like, which can be seen in Figure 3.14.

The first step is to adjust this network diagram by inserting a project buffer. The project buffer is developed using the 50–50 principle—adjusting each project task by 50 percent—and then a project buffer is added to the end of the project. This can be seen in Figure 3.15.

As shown in Figure 3.15, each task is reduced by 50 percent, and a project buffer is added to the end of the project, which totals 29 days (for an in-depth look at calculating project buffers, see Leach 2005). The estimated time to complete this project has been changed from 131 days to 95 days. Some might say that it is impossible to complete the project with such a reduction in the estimated time, and this

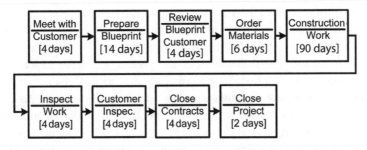

Figure 3.14 Critical chain flowchart 01

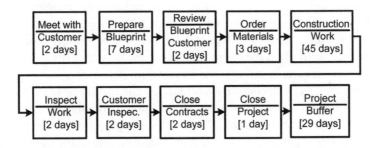

Figure 3.15 Critical chain flowchart 02

would be true if only the project plan as has been proposed thus far is changed. In order for this method to succeed, the project manager must not only change the way of preparing the schedule, but also the way the project is managed.

This is substantiated by the fact that some project managers have experienced poor project performance using the critical chain method, just as others using the critical path method have experienced project success. Good management can overcome the weaknesses of a particular method as well as make the strengths of another method futile. The key is understanding the strengths and weaknesses of the various project management methods; by knowing them, the project manager will be able to counteract the weaknesses and build on the strengths. I prefer the critical chain method because it has fewer internal weaknesses than does the critical path method. However, one key factor of performance must be changed for the critical chain method to succeed: the way that project team members approach accomplishing tasks.

The two major shifts that must take place sound rather simplistic, but they are incredibly difficult to implement, as they require a paradigm shift. First, team members must stop being required to have a project task *completed* by a certain date. Instead, they should be required to *begin working* on a task on a certain day.

For instance, if a task takes three days to complete, instead of setting a completion date, the project manager sets a start date; then the task should be completed as soon as possible, which means little to no multitasking. In this way, the project manager can remove the padding from the project schedule, which is part of developing the project schedule according to the critical chain methodology. When there is extra time available, most people will use the extra time and not turn the work in until the due date. This is because most people are hardwired from school and life in general to not complete a task until the due date, as there are no extra points for finishing early. Instead of setting a completion date, a commencement date should be set, and the task should be completed as soon as possible, which means little to no multi-tasking. If the project manager reduces the task duration estimates, as was suggested earlier, and does not change the manner in which accomplishing tasks is approached, then the project will most certainly fall behind schedule. Those who will perform the next part of the task must have some lead time as to when the baton will pass to them, so that they can start their task as soon as possible. In this way, the project team is no longer working toward a completion date, but from commencement dates, and this is one of the major keys to executing a project according to the critical chain methodology.

Another key to the success of CCPM is that the project manager should also begin systematically eliminating as much multitasking as possible from the project environment. As shown from the previous example, multitasking often lengthens the overall time required to accomplish individual tasks. Instead project managers should develop schedules that allow team members to focus on the project task at hand until that task is finished, and then move to the next assignment. Obviously, this is difficult to accomplish across the board, as individuals may be part of multiple teams with varying pressures. However, this is exactly why the elimination of multitasking must be a top-down change. Until the project manager stops pulling his team members in different directions, multitasking will lead to poor productivity. Poor multitasking will make it nearly impossible to implement the critical chain method. By eliminating poor multitasking, team members will be able to focus on one project task at a time, completing it as early as possible, and then moving to the next.

In conclusion, the critical chain method requires more than creating charts and graphs that look different; it requires the project manager to rethink her scheduling practices, estimating practices, and management practices. If implemented correctly, CCPM will lead to not only more successful projects, but also to more projects being executed successfully.

Graphical tools After a project manager has used a chosen method to create a project schedule, he may also want to create a few graphical tools that can help track construction during the execution phase of the project. One such tool is a

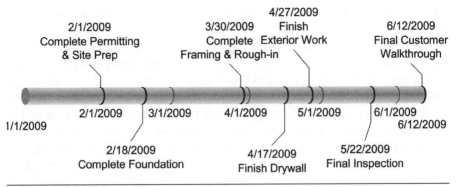

Figure 3.16 Milestone timeline

milestone chart. A milestone chart shows the scheduled start and completion time of major deliverables of the project. In this way, the project manager can see the entire schedule summarized by a few key points. Construction of the foundation, having the home under roof, hanging and finishing drywall, and the like are the types of milestones that might be placed on the chart. Many times, these milestones are points that will correspond with those points in construction that the builder is entitled to an additional payment from either the client or the bank. Milestone charts can be formatted a number of different ways, based on what information is depicted. Therefore, a format should be chosen based upon what the users will find helpful. An example of a milestone chart can be seen in Figure 3.16.

Manage the Schedule

Once construction begins, the project manager must manage not only the construction work, but also manage the construction schedule. As stated previously, the actual construction of the home will most likely not exactly match the schedule that has been created. Some factors may accelerate the schedule; others may delay it. If the schedule has a good balance, some tasks will take a little more time, some a little less and some will occur as planned. Achieving that balance is the topic of this section.

As construction progresses, the project manager will need to compare actual progress to planned progress and update the schedule as the project progresses, always preserving the original schedule as a baseline. It is to the baseline that the actual results are always compared, but the updated schedule is used to manage the project. For instance, if rain delays the foundation work for a week, the *live* schedule will need to be altered to reflect that and until the workers can catch up, all the work will be pushed back one week. Therefore, the live schedule will show dates different from the baseline schedule, as the baseline schedule will show what was originally planned.

As changes occur and the schedule is updated, the project manager must note why the delay occurred. This is important for a number of reasons. First, it provides a record for future reference. Whenever a project manger starts another project, he can review these schedule notes, which will allow a better schedule to be developed in the future. Second, it provides a record of what actually occurred in case a contract dispute arises. If the time lost was unplanned and an adequate buffer was not built into the contract, the completion date of the construction project may be pushed back, which can result in contract disputes. If these disputes were due to circumstances beyond the control of the contractor, he may not be at fault, but a record must be kept.

Recording changes to the schedule will also allow the project manager to use the recorded information to project future performance expectations. This will be discussed more fully later on, but it is sufficient to note here that by comparing actual results with planned results, the project manager can forecast whether it is likely that the project can be completed on time. If not, then she can employ various techniques to increase the rate at which the project is completed.

Cost Control Plan

Controlling costs is one of the major challenges of a construction project. Almost anyone who has spent time in the construction industry knows of a company that did not survive due to uncontrolled costs. During the boom years, builders might be able to overcome cost overruns through the volume of work done, but during average or lean years, these will be the first builders to fail. Those, however, who effectively manage construction costs will fare much better in the long run than those who do not.

How a project manager views the cost control plan will depend on the type of construction contract controlling the project. For instance, someone involved in a fixed-price contract will typically be more concerned with controlling the construction costs than someone who has a cost-plus contract. For the builder with the fixed-price contract, the lower the construction costs, the higher the profit margin. For the builder with the cost-plus contract, the higher the construction costs, the higher the profit margin. Many cost-plus contracts, however, have clauses that seek to mitigate this by offering bonuses if the cost of the project stays within certain parameters.

In most situations, regardless of the type of contract, the builder wants to control costs as much as possible. There are two basic means to accomplishing this goal: company-wide cost control policy/procedures plans and project-specific cost control plans.

Although this book is focused primarily on how to develop a project plan for one project, at this point, it is necessary to deviate a little and speak to a company-wide cost control plan. Companies will have a certain cultural approach to construction costs. This culture is typically inherited from the owner or the general manager. If this individual is a free spender, this will tend to shape the culture of the company. In this situation, the builder does not get specific quotes for a specific amount of work, but will simply agree in principle or will send out a worker without ever receiving a quote of any kind. In this situation, the bill is usually followed by surprise or anger. This type of builder will have the profit margin of projects eaten away by poor cost management.

If, on the other hand, the culture is one of frugality, most likely, the builder will seek to develop a culture of careful planning and strict cost control measures. Specific quotes on specific jobs will be acquired before work is ordered, and subcontractors will be held accountable for both performance and price. In this situation, the builder is willing to sacrifice some time in order to ensure the best value. This builder will most likely enjoy higher profit margins than the less frugal builder.

It should be obvious from the above description which model this book is advocating. The company for which I work adheres to a very strict cost control policy. The success of this company and the ability of the company to survive through difficult economic times is attributable to that fact. A company that wants to ensure long-term success should develop and foster a culture that values frugality.

The second means of controlling project costs is to develop a specific cost control plan for each project. This section will focus on the steps of that cost control plan, which are:

1. Estimate costs
2. Create a project budget
3. Employ cost control tools

Estimate Costs

The first step is to estimate project costs. During the initiation phase, the project manager prepared a preliminary cost estimate in order to submit a bid to the potential client. If the project manager employed one of the more detailed methods of estimating costs, much of the work necessary for the cost estimate is complete. If a ballpark price was given then, the project manager will have to perform a detailed cost estimate now. This section will assume that a ballpark figure was given. Although a number of methods to prepare a cost estimate are possible, this book will follow a modified form of the bottom-up estimating method. Entire books have been written on how to estimate construction costs. Therefore, this brief overview will offer the bare basic tools necessary to perform an estimate.

In order to prepare this cost estimate, the project manager will need the materials that have been previously prepared, which are:

- Project scope statement
- WBS
- WBS dictionary
- Construction schedule
- Estimated task resources and duration

In a true bottom-up estimate, the different parts of the project are broken down into smaller components; next, a cost is assigned to each of these smaller components; then they are aggregated together to provide a quote for a certain portion of the project. After all these small components are assigned a cost and aggregated together, the project manager simply adds them up to obtain a total construction cost estimate. A full bottom-up estimate may be necessary at times, but most of the time a partial bottom-up estimate will be sufficient, especially if much of the work is subcontracted out. For those who rely much more heavily on in-house employees, a true bottom-up estimate may be necessary, but most builders in the residential construction industry rely heavily on subcontractors.

The partial bottom-up process contains these steps:

1. Develop a list of tasks to estimate
2. Develop a description of the work to have estimated
3. Request labor quotes from subcontractors
4. Request material quotes from suppliers
5. Develop estimates for any in-house work
6. Combine gathered material into a completed quote

The list of tasks to estimate will closely resemble those tasks listed in the WBS. Some project managers will be tempted, however, to combine phases of certain tasks together, such as combining the rough-in plumbing and finish plumbing into a simple plumbing line item. It is important to not do this for reasons that will become clear later on. Tasks—such as plumbing, electrical, insulation, and the heating and cooling systems—that have more than one phase should be kept separate in the quote, as it will allow for better tracking and projection of costs later on.

A sample task list is:

1. Initial site grading
2. Footing
3. Foundation

4. Basement slab
5. Termite pretreat
6. Framing
7. Roof
8. And so on

Once the task list has been created, the project manager then gathers the information or materials that will need to be sent to the subcontractors to request quotes. Most of this information should be found in the WBS dictionary and the information developed for the project schedule. If the subcontractor is expected to provide the materials, that should be stated in the request for a quote. One might even request that the quote be broken down into materials and labor; this allows the project manager to determine whether the materials can be secured by the company for a lower cost. If this route is chosen, the project manager must make certain that materials of a similar grade and quality are being quoted. Once all this has been determined, the requests for quotes should be sent to the subcontractors. After the quotes have been received, the project manager must review them to make certain that the subcontractor understood the request and has quoted appropriately.

For those materials that the builder will be providing, the project manager will have to gather quotes from suppliers. Builders will approach this from different perspectives. Some builders, not wanting to deal with multiple suppliers, will choose to get as many materials as possible from one or two suppliers. This greatly reduces the time in gathering quotes, and it makes the quotes and the suppliers easier to manage. Other builders, such as the company for which I work, choose to take the time to get quotes from a variety of suppliers for different products in order to take advantage of the best prices. In this practice, quotes should be sent to multiple lumber yards, concrete companies, electrical and plumbing supply companies, and so on, in order to gain the best pricing. Once the project manager knows who the most competitive suppliers are, he can approach those suppliers first in the future for supplies, then periodically check their prices with other suppliers to make certain they are still the most competitive.

Such an effort may seem foolish to some, but it is a matter of perspective. Consider this example: Supplier A provided a specific service for $850 per home. Supplier B offered the same service for $550 per home. The difference of $300 might seem like an insignificant sum to some builders, but consider the fact that this service is performed on an average of 40 homes per year. Therefore, the savings by switching providers is approximately $12,000 per year, which is a substantial sum for most residential builders. By saving a few hundred dollars per job on a variety of portions, the builder can significantly increase the bottom line.

The next step is to develop any quotes for in-house work. If direct employees are going to be performing a certain portion of the work, the value of their labor needs to be assigned a cost rate. For instance, let's say a company performs all site grading, installs the footings, installs the septic tanks, and a variety of other tasks that are not subcontracted out. A certain value is assigned to each of these tasks and included as part of the cost estimate. Over time, the project manager will develop something of a standard rate to assign for these various tasks, but if she is uncertain, then she must calculate the time, people, machinery, and the materials required and add them together to come up with a total cost estimate, which is the purpose of the next step: create a project budget.

Create Project Budget

The project budget is created by aggregating all the individual quotes into one comprehensive document, which creates a total cost baseline for measuring project performance and progression. Once all the individual quotes have been combined, they need to be compared with what was initially quoted either to the customer or to the internal project sponsor. If any of the cost estimates come back vastly different from what was expected, then it may be time to have a meeting with the appropriate stakeholders. For instance, if the builder is involved in a cost-plus contract and the initial estimated cost was $250,000, but now the more detailed quote is $325,000, it is time to have a meeting to discuss what has changed, why it has changed, and what is going to be done about it. This is why it is important to provide detailed and reliable quotes during the initiation phase of the project. If this kind of budget change were to happen on an internally initiated project, then the project might be cancelled or modified to meet the financial constraints or goals of the company. The important thing is that if a red flag is raised, it is important to deal with it earlier rather than later.

After the costs have been aggregated together into a cost baseline, a reserve fund or buffer should be created as a way to plan for the unplanned. There may not be enough information to do this yet, as the risk management plan has not been completed yet. So this step may have to be revisited later on, but it is important to note that unexpected costs will arise. These unexpected costs must be appropriated within the budget.

The next step in creating the project budget is to determine funding needs throughout the life of the project. This is done by combining the cost estimates and the construction schedule. A sample budget time line is shown in Table 3.1.

This type of report allows the project manager to see when funds will be needed either from the customer or from the company, which helps ensure that the necessary funds are on hand as necessary. Once the report is prepared, it should be reviewed with the appropriate stakeholders, such as the customer or

Table 3.1 Budget time line

Task	Date to be completed	Budgeted amount	Aggregate amount
Site grading	May 1	$2500	$2500
Foundation	May 8	$4500	$7000
Framing	May 16	$7000	$14,000
Rough-ins	May 27	$9000	$23,000
Insulation	June 1	$3000	$26,000
Drywall	June 14	$4500	$30,500

individual responsible within the company. The example shown in Table 3.1 is not detailed due to the space constraints. The project manager will not have a space constraint and so will probably want to prepare a much more detailed budget time line. The project manager can prepare one time line for costs associated with labor and one with material costs, or the two may be combined as they are in the example. The important part is to provide enough detail to be helpful, while not being burdensome.

If this is an externally initiated project with a client, then it is likely that he will have a construction loan for the project. In this case, the builder must review the budget with the bank to make certain that the bank's payment schedule will align with the payment schedule developed internally. If the two schedules are not compatible, then some type of plan will have to be negotiated to make certain that funds are available to pay for the construction project. Most of the time, this is not a problem, but it should not be taken for granted.

So far the cost control plan includes a detailed cost estimate, a project budget (cost baseline), a contingency fund estimate, and a funding schedule. The last step is to outline how changes in cost will be handled and what control analysis techniques will be used to track costs as the project progresses.

During the time that the project progresses, it is likely that the costs of some materials will change. Or it may be that the client or the builder decides to change some features in the home, which requires new estimates. Regardless of the cause, it is important to know when and how to change the project's budget to reflect these new needs.

The project manager must keep in mind the purpose of the project's cost baseline when considering the modification of the project's budget, which will also modify the project's cost baseline. The cost baseline is a record of planned and expected expenditures. It is used to compare actual results with estimated results to determine how the project is progressing. It will help the project manager determine if her estimating was accurate. It will be used to forecast the final cost of the project during the construction phase, as well. At the conclusion of the project, the baseline will tell the project manager whether the actual results match the planned results. If not, then she uses the project baseline as a means to determine what went wrong, when it went wrong, how it went wrong, and how to plan better in the future. Therefore, the cost baseline should not be changed each time there is a minor adjustment to the project.

Given the importance of the cost baseline, changing it is a major consideration that should receive proper authorization. Typically, the project baseline should only be changed when there is a major modification to the project plan that justifies such an action. Consider a couple of examples:

> *Example 1: The quote received for the framing materials expires before the materials are ordered, which results in a 5 percent cost increase for the framing materials. Should the cost baseline be modified? No, the change in material cost does not justify a modification to the project's cost baseline. As the project is analyzed, the project manager will see the variance and will investigate the cause. If someone did not place the order on time, the issue should be addressed with that employee. Additionally, if the project is behind schedule, it should be notated.*

> *Example 2: The customer decides to add an in-ground pool to the home during the middle of the construction project. In this instance, the change will be handled according to the integrated change management control system. This type of change will definitely affect the cost baseline, as the pool is a substantial change to the project. If it is not added, then it will appear that the project has been delayed and the cost has increased for no cause. Therefore, by modifying the project baseline, the project manager ensures that the actual results can be compared to the planned results in a way that yields profitable information.*

Because the project's cost baseline should only be modified in certain situations, it is important to set the means by which it can be modified. Typically, it should require the authorization of the project manager, an officer of the construction company, and possibly the client. A simple form that states the change and the reason for the change and provides authorized signatures for the project file can be developed. As a side note, the project manager should ensure that the client can actually afford the change requested before modifying the project and implementing it.

Employ Cost Control Tools

The last part of the cost control plan is to determine which, if any, metrics will be used to analyze the project's cost performance. There are a variety of tools available, but only a couple of the more simple ones will be considered here; variance analysis and earned value analysis.

In a variance analysis, the project manager simply compares the actual results with the planned results in a spreadsheet or graph format to demonstrate how much variance there is between the actual results and the planned results. The analysis can focus on cost or time or both, as is shown in Table 3.2.

In this table, the budgeted cost of the various tasks is compared to the actual costs of each task, and a percentage is given to reflect the difference. Some might want to insert a dollar amount in the variance column instead of the percentage shown in Table 3.2. I prefer the percentage because it shows the impact relative to the amount. For instance, a $200 cost overrun might not grab someone's attention, unless they realized that the planned cost was $100. This would be a 200 percent overrun, which would catch my eye quicker than $200. All this is of course relative to the size of the project, but I do find percentages to be more helpful than dollar amounts in this report.

Many project management software packages will create this type of spreadsheet or chart with a few clicks. The results, however, are only as good as the information that has been entered. If incorrect or flawed data is entered, then the charts will be incorrect or flawed. Microsoft Excel© or Microsoft Works© can also produce spreadsheets and charts.

A variance analysis is beneficial because it is simple to do, and the data, presented properly, make it easy to identify areas that need to be addressed. Although a variance analysis cannot tell why there is a variance, it can point out where the variance is. The project manager must then investigate why the actual results do not match the planned results.

An earned value analysis is a slightly more technical method of analyzing cost data. It is also referred to as EVM (earned value management). EVM is a project performance measurement and monitoring tool, as well as a forecasting tool. It objectively depicts the relationship between scope, schedule, and resources in a particular project in such a manner as to show the project manager how well the project is performing now and is expected to perform in the future. It shows these relationships by the use of some rather simple mathematical ratios or formulas.

Table 3.2 Variance analysis

Task	Budgeted amount	Actual amount	Variance	Budgeted days	Actual days	Variance
Site grading	$2500	$2300	-8.00%	4	4	0.00%
Foundation	$4500	$4200	-6.67%	7	6	-14.29%
Framing	$7000	$7500	7.14%	8	10	25.00%
Rough-ins	$9000	$9000	0.00%	9	8	-11.11%
Insulation	$3000	$2800	-6.67%	4	4	0.00%
Drywall	$4500	$4900	8.89%	13	14	7.69%

The benefits and value of using EVM are many. First, it provides a universal way to view project performance. Many times builders need a way to determine which project to focus the most attention on or which project is falling behind schedule or is going over budget. By using the proven quantitative tools of EVM, project managers can compare projects more easily. Those projects that are beginning to slip in one area or another can be isolated, and the necessary attention can be given to those projects.

Another benefit of EVM is that it will also allow the builder to estimate the future project performance based on past performance. It is easy for those within the construction industry, like those within other industries, to be primarily reactive rather than proactive. They assume that they have no way of forecasting what the future will hold for their project, but EVM provides a solution to this problem. Granted EVM will not tell them if the price of lumber or concrete will increase, but it will tell them with relative certainty if their project is going to go over budget. This could be very advantageous to builders who are working under fixed-priced as well as cost-plus contracts. For those working under fixed-priced contracts, EVM shows that their profit margin could be decreasing by the day. For those with cost-plus contracts, it alerts them to the fact that the future homeowner will be unpleasantly surprised. But most importantly, it alerts the builder that some corrective action, if possible, needs to take place, which brings us to one of the weaknesses of EVM.

Once a project manager has forecasted either a schedule overrun or a negative cost variance or both, some action must be taken to determine how the project can be brought back into control. EVM does not tell the builder what action needs to be taken, it merely alerts the builder to the fact that an undesirable variance has occurred, which if unchecked could spell ruin for the project. It is important for the builder to realize that some variance is to be expected—maybe some small percentage of float between the negative and the positive—but if the trend is continually moving toward the negative side, some action must be taken. Herein lies the EVM practitioner's quandary: EVM does not tell the project manager what to do to correct the variance. It only reports the relationship between what actually happened and what was planned. The project manager must determine what corrective action to take. As within any performance measurement tool, the human element is critical, which is why proper training is imperative to EVM success.

Here's how EVM actually works. In the simplest terms, one takes the actual performance and compares it to the planned performance using certain formulas. The results of these formulas will inform the project manager if the project is over budget, under budget, ahead of schedule, behind schedule, and so on. A detailed example of how to perform this type of analysis is provided in a later chapter which covers controlling the project.

Whether a project manager uses EVM or variance analysis, the important point is that he is comparing the actual results with the planned results in such a way that he can track the progress of the project. Now that the cost control plan has been covered, it is time to focus on the quality assurance plan.

Quality Assurance Plan

The quality assurance plan is focused on two items. First, it is focused on the quality of the actual work. It is imperative that the project manager plan and execute those steps necessary to make certain that the home being built meets the expected level of quality. Take note that it should meet the *expected* level of quality. This implies that there is an understanding that exists between the key stakeholders as to what the quality of the structure should be. This means that not only is a specific level of quality planned for, but it is also evaluated as construction progresses.

Second, the quality assurance plan focuses on ensuring that the project is performing according to plan. Developing the plan is a major undertaking. But consider that someone takes the time to develop a plan, as has been discussed and will be discussed, but the time is never taken to ensure that the plan is followed. It is never verified that the budget is being followed, or that the schedule is being updated, or that the potential project risks are being watched, or that the change orders are being properly recorded and administered. This would no doubt lead to the failure of the project. Therefore, the quality assurance plan also specifies how to ensure the quality of the project itself.

The quality of both the construction work and the project work is planned and documented in the quality assurance plan. First, the quality of the construction work will be considered, then attention will turn to the quality of the management of the project.

Establish Quality Standard

The first step to developing a quality control plan is to determine what the standard of quality is. It is important to note that quality is a relative term depending on a variety of factors. For instance, the materials and methods that produce what is considered a well-built quality home in one section of the country might be considered inadequate in another area of the country, or even the same state. Because of this, this book will not focus on developing specific quality standards, but will instead focus on how a project manager goes about establishing specific quality standards and monitoring adherence to those standards.

The following is a list of steps to follow to set a quality baseline:

1. Know the target audience's expectations
2. Become familiar with local building codes and inspector's expectations
3. Become familiar with those best practices used by experienced builders
4. Question subcontractors
5. Review industry publications

The first step to establishing the quality baseline for a construction project is to have an idea of the customer's expectations. If there is not a specific client yet, then the project manager should learn more about the type of client she might expect to purchase the home being constructed. Different customers have different expectations. The individual who is seeking to purchase a 1500 square foot home for less than $200,000 has expectations different from someone who is seeking a 1500 square foot retirement home for $400,000. This person will expect different features, different materials, and different levels of quality, just as a person purchasing an affordable car expects a level of quality different from someone purchasing an exotic import might expect. The key is to learn what the expectations are. This is relatively easy if the builder already has a client in hand and is able to discuss the issues with him. Remember that a client's expectations may go against the local building codes or even best practices. In this instance, the builder must educate the client in a gracious way.

The second step to establishing a quality baseline is to become familiar with local building codes and the expectations of local building inspectors. It is not expected that a project manager will immediately memorize all the details of local codes; the goal is to become generally aware of the guidelines and be able to reference them at the appropriate time. I can recall reviewing these documents while taking a prep course for the residential contractor's license. The instructor, a veteran builder, was covering the *code house*, and he made the statement that no one wants a code house. By this he meant that the building code is a list of minimal requirements, not necessarily best practices.

After a project manager becomes familiar with the applicable local building codes, he will need to begin developing relationships with the local building inspectors. They typically hold office hours during which one can go by and meet them and discuss various issues. This is a good opportunity for a project manager to develop a cordial and professional relationship. The goal is not to ask for favors or for exceptions, but to learn about how the inspectors go about interpreting the building code. Two inspectors can look at the same building code and have slightly different expectations about what should be done to comply with the code. Therefore, if an area is unclear, they should be questioned. Also, a project manager might consider asking them what areas in construction they see as the greatest challenge; this is a clue to what they will pay particular attention. Unfortunately, many people see building inspectors as the opposition. This should not be the case. In general, building inspectors are professional individuals who perform a vital function.

The third step to develop a quality baseline is to become familiar with the best practices used by seasoned and successful builders. One of the best means of accessing these individuals is through the local home builder's association. Typically, a meeting is held each month where a speaker will discuss some aspect of building or speak about current trends. The purpose of this network is to develop relationships within the industry and help one another succeed. Many people are surprised at the amount of help that is available through such an organization.

Another excellent source of best practices is the local community college. At many community colleges, a number of seminars and classes are offered on a variety of topics: framing, green building, flooring, etc. By taking advantage of these opportunities, a project manager will learn helpful information and continue to grow her business network.

A fourth method of gathering information to establish a quality baseline is to question subcontractors. When subcontractors are given the plans to review, the project manager can ask them how they will do the job, what options are available, and why they recommend one method over another. She is likely to hear different opinions from different subcontractors so the information must be processed in a discerning manner. Many times, subcontractors will describe methods or provide useful hints that provide insight and prove to be useful. It is important to listen, not act as though what they are saying is already known.

A final suggested method is to review industry publications. Most of the various portions of the construction project represent entire industries, such as the electrical, plumbing, HVAC, and concrete industries. Most of these industries have official publications that describe best practices, quality standards, and an abundance of other information that can be highly valuable. A project manager should take the time to subscribe to these types of publications and review them often to stay up to date on developments within the field.

The methods described above are ways that a project manager can develop a knowledge of and stay updated on quality standards for various markets. Some may be aware of other methods, and if so, they should take advantage of those. The point is to provide some practical instructions on how to develop an awareness of those quality standards that should be followed to construct a well-built home for a local market.

The second quality baseline that needs to be developed relates to how the project will be managed. Books such as this offer best practices for managing residential construction projects. There are also a host of other resources available: other books, seminars, construction project management classes, and industry magazines to name a few. These resources provide information on how a project manager should either manage the entire project or certain aspects of the project. By becoming aware of the various management programs and techniques available, he can choose those that are most applicable to each situation.

Once the project manager has an idea of what constitutes a well-built home and what project management methods are helpful, it is time to develop a quality management plan for the project. The quality management plan simply spells out what has been learned and applies it specifically to the construction project at hand. In order to effectively apply the learned information to the project plan, the project manager will need to be discerning. Not everything learned about constructing quality homes will be applicable to every project. So he will need to determine which quality control measures are appropriate for each specific project.

Select Quality Control Measures

Determining which quality control measures are appropriate is called quality planning. Think of it this way. In a skilled carpenter's tool box are a number of tools—some are general tools and some are very specific in their use. Whenever the carpenter takes on a job, the job must be analyzed to see which tools are necessary. If the carpenter brings the wrong tools to the job, then he will be unable to complete the job. If he brings all the tools, then he has most likely wasted time and resources. This simple example parallels the task of planning quality. The experienced construction project manager knows a number of methods to construct a home, but also a number of techniques to manage a project. The project manager must determine which tools are most useful for the job at hand.

The purpose of quality planning is to develop a quality standard for the actual construction work. Thankfully, the project manager should have a detailed set of plans and specifications that detail how the home should be constructed. She should review these plans with a discerning eye and ask pertinent questions to any support persons (engineers, architects, etc.). But in general, she is being told the level of structural quality to which the home is to adhere. What she must do is ensure that the work is done according to the specifications.

Therefore, in planning the quality of the actual building, the project manager will want to review the plans, specifications, and project schedule to ensure that inspection points are inserted to ensure that the work complies with the appropriate guidelines. If the project manager is unable to do this, someone who is knowledgeable must be assigned—someone who would recognize a problem if one exists.

The second phase of quality planning is to develop the means of ensuring that the project performs as planned. How does a project manager ensure that materials arrive on site, for example, that the schedules are updated as work progresses, that the contracts are properly administered, etc.? This is the purpose of quality planning. Throughout this chapter so far, only a couple of sections of the construction management plan have been discussed. As each new section of the plan is developed, it may be necessary to go back to update sections with new information. Almost every section will have to be updated after the quality planning is completed, so that everyone knows how to ensure that the project is being performed according to plan.

One of the best methods of ensuring that everything is done according to plan is a simple one: checklists. When I first graduated from college, I decided I wanted to take airplane flying lessons. Having ridden in my uncle's small twin-engine plane, I had become fascinated with flight. I almost immediately began taking lessons at a local airport. On the first day to go up, I learned an interesting thing. I would spend almost as much time on the ground going through inspection checklists, as I would in the air learning to actually fly.

Those checklists were simple tools, but they were vitally important. If I were to disregard those checklists, I would risk not simply the airplane, but my own life and the life of any passenger aboard. Developing and using checklists can be rather tedious, but they are great tools to ensure that work is done properly. Risk management checklists, cost management checklists, quality control checklists, contract management checklists, scheduling checklists, and a host of others can be developed to ensure that the project is being managed properly.

After a project manager has determined when he will inspect the construction project and how he will ensure the quality of the project plan, he needs to perform a quality audit. The quality audit inspects the construction plans and the construction management plan to ensure that the appropriate quality measures are in place. Failing to do this can create major problems. Early on in my construction career, I participated as a minor stakeholder in a particular internally initiated construction project. The company was building in a new area and was unfamiliar with the building inspectors, as well as the local best practices. The company was going to build a new home style in an established neighborhood, which required the approval of the home owner's association's architectural review board (ARB), as well as the approval of the local building inspection office. The plans were first submitted to the ARB, as it was a fairly straightforward and free process. The ARB was primarily concerned with aesthetics, not structural issues. The home was approved without any problems arising. Next, the plans and construction specifications were sent to the local building inspection office for review. About two weeks later, the plans were returned and were marked "approved." Construction got underway. The home passed the footing inspection, the foundation inspection, the basement slab inspection, but failed the framing inspection. The framing inspector stated that the rear foundation wall lacked the structural integrity necessary for the wind zone in which the home was being built. The plans were reviewed and compared to the actual work, and it was discovered that the actual work matched the planned work. Therefore, the home complied with what had initially been approved by the building inspection office. However, the construction guidelines were inappropriate for the wind zone. The building inspector who reviewed the plan did not check them thoroughly, signed off on them, and returned them approved. The project manager did not think anything about it, because the plans were approved by the building inspector. The point is that if a quality audit had been thoroughly performed, it is likely that this mistake would have been caught, as the quality audit seeks to ensure that the proposed construction project follows not only the best construction practices, but also meets the local building codes. Performing the necessary quality planning during this phase of the project will save much time and money later on.

So far, the discussion has focused on developing quality standards and developing tools to manage the quality of the project (checklists). The last part of the quality assurance plan deals with actually performing quality control on the construction project.

Perform Quality Control Measures

As the project progresses, what is the plan to gather quality control information, how are defects handled, what is an unacceptable level of defects, and how will repairs be approved? A final step must be taken to develop a plan to implement the quality assurance program. Some larger construction companies have quality control officers who are loaned to projects on an as-needed basis to implement quality control measures. This is not the case for most construction companies. Therefore, the project manager must assign a trusted individual to perform this vital task.

As the project progresses, the individuals responsible for quality control must use their knowledge, the information provided in the construction specifications and plans, and the quality checklists to ensure that the project adheres to the plan. They will perform inspections at predetermined times and will report the findings to the project manager and interested stakeholders. These individuals are not merely to identify problems, but they are also responsible for identifying the sources of the problems. In this instance, they are to act as investigators, trying to determine what gave rise to the undesirable result. Was it a process failure, a machine failure, a human error? This is not to be the proverbial witch hunt, but it should be an earnest desire to learn how the root cause can be effectively handled.

There are a number of tools that the quality control team can use to help with this task: control charts, flowcharting, cause-and-effect diagrams, brainstorming analysis, and inspections. The two primary methods advocated in this book are cause-and-effect diagrams and structured brainstorming.

One of the most effective methods for construction projects is cause-and-effect diagrams, which are also called Ishikawa or fish bone diagrams. The cause-and-effect diagram shows how various factors have influenced a particular situation. The other method, structured brainstorming, is a method in which people are brought together and are led in a session to both explore and explain a specific issue. These two methods work very well together, as shall be seen in the following example.

A construction delay is one of the most common problems construction project managers face. Imagine that a nine-month construction project is four months underway and is considered to be one month behind schedule. First, there is the question of how this was determined. This could be effectively determined by looking at the construction schedule and performing some earned value management techniques. But just assume for the sake of the example that the estimated delay is correct. If the project is going to get back on track, then it must be determined what caused this delay. It could be for an obvious reason such as a hurricane or flood or some other event outside of the control of the project manager, but this is not the case for most construction delays. Most construction projects are not delayed because of one major event, but because a number of minor delays go unnoticed and uncorrected until it is almost too late. Suddenly the team finally wakes up and realizes that if the problem is not corrected, the project will not be completed on schedule.

A cause-and-effect diagram is one means of learning the cause of a problem. There are a few basic steps in constructing a cause-and-effect diagram:

1. Identify the problem
2. Identify contributing factors
3. Identify cause(s)

First, the problem or the visible effect of the problem has to be identified and agreed upon. In the case of this example, the problem is that the project is behind schedule. It should be fairly easy to gain consensus on the fact that the project is behind schedule. Then once the problem has been identified, it is placed in the cause-and-effect diagram, as shown in Figure 3.17.

Now that the problem has been identified, the next step is to list and categorize contributing factors. It is possible that there is one factor that has led to the schedule delay, but it is more likely that there are a few contributing factors. This is where

Figure 3.17 Cause and effect diagram Phase 1

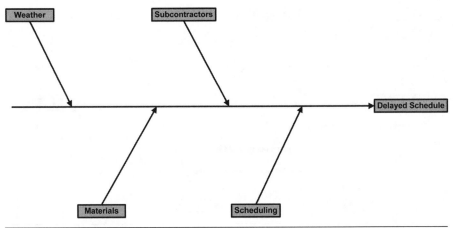

Figure 3.18 Cause and effect diagram Phase 2

structured brainstorming can be useful. It is useful to work backward with the project schedule. For instance, the project manager can start with the most recent task and ask why it was not started or completed on time. The reason might be that the subcontractors or materials did not arrive on time, or that the weather impeded the work, or that the previous task was not completed on time. The project manager lists the possible reasons and moves to the previous task on the project schedule. This is repeated until she arrives at the point in the project schedule at which the delays first began to appear. Once the possible reasons for these delays are listed, the reasons are categorized according to type, such as subcontractor related, vendor related, communication related, weather related, and so on. The categories are added to the cause-and-effect diagram, as shown in Figure 3.18.

At this point, one can see the types of issues that led to the project's being behind schedule. This helps everyone to think categorically. If the problems are categorized, it is easier to think through who should have been responsible for correcting the problem or preventing the problem. Now is the time to revisit the brainstorm list and further refine and define specific issues according to the categories that have been inserted into the diagram. The result might appear as shown in Figure 3.19.

Now the cause-and-effect diagram has served its purpose: identifying those issues that led to the problem. Some cause-and-effect diagrams will be more helpful than others. The more specific they are, the more helpful they will be. However, this is where one reaches the limit of the cause-and-effect diagram and structured brainstorming—these tools only identify the problems. Now the project manager must develop solutions to the problems. How can these problems be corrected and

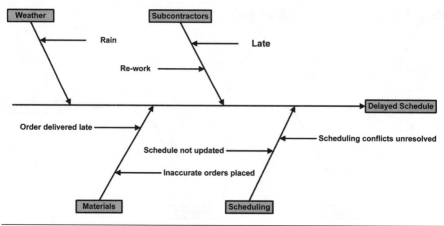

Figure 3.19 Cause and effect diagram Phase 3

the project get back on schedule? The project team will need to develop specific solutions. Some of the possibilities will be discussed later in the risk management section of the project plan, but the ability to solve problems is what separates successful project managers from the less successful. All sorts of methods can be developed to identify problems, but correcting the problem requires the knowledge and expertise of the project team members.

Human Resource Plan

The human resource plan is primarily concerned with how people will be hired, how they will be developed or trained, if necessary, and how they will be managed. This applies to not only employees of the construction company, but any subcontractors that might be hired to work on the project. Much of residential construction is outsourced to subcontractors. Therefore, project managers must have a definitive approach to acquiring, developing, and managing subcontractor teams.

A construction project will hire primarily two types of people: internal employees and subcontractors. Internal employees are full-time or part-time employees of the construction company. Subcontractors refer to those companies or individuals hired from outside the construction company to perform a specific task on the project. So the human resource plan will deal with both types of people.

The human resource plan focuses on the following steps:

1. Determine staffing needs
2. Acquire staff
3. Develop the timetable
4. Identify training needs
5. Offer recognition and rewards
6. Ensure compliance
7. Ensure safety

Determine Staffing Needs

The first step is to determine the staffing needs of the project. This includes those who do the actual construction work, as well as those who perform administration. The administrative team is responsible for paying bills, placing orders, calling vendors, updating reports, managing contracts, and a host of other activities that are necessary for a home to be constructed. On smaller projects, this team might comprise only one or two people, but the work is very important. So both aspects of the project need to be considered.

The WBS and the WBS dictionary will inform the project manager as to staffing needs for the construction work. He can review the types of work to be done and determine what types of skilled laborers are necessary for actually constructing the home. A list should be created that details the types of workers needed and the number desired. Then the project manager should look at the in-house employees to see who is available to do which jobs. In residential construction, many of these jobs will be outsourced to subcontractors. Sometimes the company may even have an employee who can do the work but still might choose to outsource the work because of scheduling conflicts or any number of reasons. If the construction plans call for a specific task that the project manager is unfamiliar with, he may need to consult a subject matter expert or an engineering firm for guidance.

A second list should be developed that describes the administrative side of the project. This list would include jobs such as accounts payable, purchasing, scheduling, stakeholder management, and the like. Some of these tasks will most likely be done by general company employees who are not assigned to specific projects, but act in support roles for all construction projects. If this is the case, the project manager will need to develop a plan for making certain that these individuals have the necessary tools and information to perform their support roles at the appointed time. If the project manager does not effectively manage these support roles, the project may develop time delays or cost overruns. It is her responsibility to make certain that these people know their jobs and do them effectively.

Acquire Staff

The next step is to develop a plan for acquiring the project team. For the construction side of the project, this means first determining which tasks will be performed by internal employees. After this has been done, the project manager knows how many subcontractors will need to be hired to work on the project. Most companies already know which subcontractors they will use, because they have developed relationships with the subcontractors in the past. If not, the project manager will have to contact subcontractors and request bids and discuss availability. Once the project manager knows who will be assigned internally and what positions will require subcontractors, he must develop a timetable for acquiring internal employees and subcontractors.

Develop Staffing Timetable

Developing a timetable sets out a plan for dates by which a resource should be acquired. Depending on the size of the company, internal employees are typically easy to access. Simply make a request or tell the employees that they will be working on this project at this point in the future. In this instance, the project manager will need to follow whatever protocol the company has established.

The project manager also needs to develop a timetable for acquiring subcontractors. One strategy is to set aside a section of time in which to try to locate all the necessary subcontractors. More likely, however, the project manager will develop a schedule by which certain subcontractors should be located. This reduces the amount of work required up front so that the project can possibly begin more quickly. It is important to locate a site-grading subcontractor before locating a framing crew. The point is that if the project manager has a plan in place to locate the right subcontractor at the right time, then it should not be necessary to locate everyone before the project can begin. However, this will depend on each project manager's preference as there are pros and cons to each method.

Identify Training Needs

After a plan has been developed to acquire the project team, the project manager needs to determine whether it will be necessary to offer training to any participants in the project. Most likely this would apply only to internal employees and not subcontractors. If a project manager fails to make certain that the employee's skills match the assigned task, then she should not be surprised when the work has to be redone. The general assumption is that a subcontractor is capable of doing the work for which he has bid or he would not have submitted a bid. But the project manager should take reasonable steps to ensure that the subcontractors are competent and capable to perform the tasks on which they have bid.

If a project team member lacks the skills to perform an assigned task, then the project manager must decide whether she should provide training or whether she should assign the task to someone else. In this instance, it is useful to perform a simple cost-benefit analysis. The project manager simply compares the time and resources necessary to train the employee with the benefit of doing so. If the benefits outweigh the costs, he should probably offer the training. If not, he should assign someone else.

Offer Recognition and Rewards

Next, the project manager should develop a system of recognition and rewards for good performance. For internal employees, this might mean a positive review, which might result in a promotion, a bonus, or a pay increase. For subcontractors, it might include giving them preference in future contracts or giving them a bonus. People like to be recognized when they have performed their jobs well, and they like to be rewarded. The reward does not have to be large or extravagant, but it should be appropriate. It should also not be arbitrary. The project manager should carefully think through a reward system and honor all people equitably.

Ensure Compliance

The human resource plan should also include assurances that the project is being operated in compliance with all governing bodies, union contracts, and company human resource policies. Discussing the specifics of this is beyond the scope of this book, but at a minimum it is the project manager's responsibility to ensure that no laws are broken when dealing with the human resources aspects of the project.

Ensure Safety

The last part of the human resource plan involves job site safety. How is the project manager going to ensure the safety of all workers on and off the job site? Once again, the specifics of this are beyond the scope of this book, but it is imperative that the project manager take safety concerns seriously and take the steps necessary to create a safe work environment.

Communication Plan

One of the most important aspects of any project is communication; this is especially true with construction projects. The communication plan will provide a comprehensive guide to making certain that employees, subcontractors, vendors, customers, and other stakeholders receive communication in a timely and accessible manner. The words *timely* and *accessible* have taken on new meaning in the past decade or so. As a child, I can recall my father using a citizens band (CB) radio to communicate with the home office while he was on the job site; then came the infamous bag phone, which he would carry with him on the job site. He stopped upgrading within the past few years, not wanting to E-mail or text from his cell phone, but the transformation since the 1990s has been somewhat amazing as the communication methods have continued to change.

There are a number of tools that can be employed to communicate with the project team and stakeholders: cell phone, fax, text messages, E-mail, even Twitter. Some builders are seeking to leverage new communication tools, such as Facebook, Twitter, private message boards, and blogs to enhance communication for the construction project. All the more old-fashioned tools (telephone, fax, E-mail) are useful and so are the newer methods (Twitter, Facebook, text messaging), but the primary question is what communication methods will most effectively meet the needs of the current project.

Another issue that must be considered as the methods of communication increase is contract law. Those involved in construction management may be parties to dozens of contracts at any given time. These legal contracts most likely contain deadlines for communicating various types of information to various people at certain points and times. Does a *tweet* (a short message) on Twitter qualify as a legally valid form of communication? These are some of the issues that must be considered when creating a project communication plan.

The communication plan answers the following questions:

1. Who will update the communication's plan?
2. Who needs information?
3. What information is needed?
4. When is information needed?
5. How should information be sent?
6. Who is responsible for sending information?

Who Will Update the Communication Plan?

The first point to address is who is responsible for the communication plan. This does not refer specifically to the person(s) who will create the plan, but rather who the information officer for the project will be. For most residential projects, this will be the project manager, and it will be her responsibility to keep the plan up to date as the project progresses. If the plan is not kept up to date, then it is only a matter of time before the project begins to fall behind schedule, cost more than expected, or become subject to any number of other potential problems.

Who Needs Information?

The second point to address is who needs information. The list may be quite long for most residential construction projects. Three excellent sources of names will be the project stakeholder list, the project team roster, and the scheduling documentation, which lists those who will be working on the project. At this point, the project manager may not have hired all the specific subcontractors, so he may need to substitute generic names such as *plumbing subcontractor* in place of a specific company name. This will have to be updated once the specific subcontractor is hired. The project manager will want to think through this list carefully, as it may need to contain people that he might not initially include, such as those who live near the construction site. These individuals' lives will be disrupted in some way for the next few months. Providing them with updates and attempting to placate them in some small ways may prove very beneficial not only for the success of the project, but for the reputation of the company. Once a comprehensive list has been developed, one is ready to move to the next step, which is determining what information all these people need.

What Information Is Needed?

Determining what information people need may seem like a tedious task, especially if the project manager is the one who is doing most of the communicating. After all, if the project manager knows who needs what, then why must it be written down? It should be written down, because this ensures that the plan has been thought through and that there is a record in case the project manager is incapacitated or if a legal dispute arises. The plan should list all of the major communications that should be sent to the various stakeholders. Consider the following examples:

Communication to framing subcontractor:
- Request for bid
- Copy of construction blueprints and specifications
- Copy of materials list
- Approval of bid
- Subcontractor agreement contract
- Schedule of work (all updates)
- Construction updates
- Inspection reports during and after completion of framing
- Approval of work completed
- Payment for work completed
- Closing of subcontractor agreement contract

Communications to building inspector:
- Application for building permit
- Copy of all plans and specifications
- Requests for property inspections and reinspections
  - Footing inspection
  - Foundation inspection
  - Slab inspection
  - Framing and window inspection
  - Rough-in inspections (electrical, plumbing, HVAC)
  - Insulation inspection
  - Prefinal inspection
  - Final inspection

Communications to job site neighbor Mr. and Mrs. Smith:
- Notification of construction project
- Request if they have any special conditions or accommodations that need to be considered
- Monthly summary status reports
- Notifications of any early (pre–7AM) or late (post–6PM) work

The lists above show how different types of stakeholders will receive different items of information as the project progresses. The first time that someone creates a communication plan, it is quite a bit of work, but once it is created it can simply be adapted to future projects, as most construction projects have the same types of stakeholders involved. Taking the time to do this will help make certain that the right information is distributed to the right people.

When Is Information Needed?

The next step is to determine when the people will need the information that has been listed for each of them. This can be a bit tricky to think through as a project manager's first inclination is to assign a date based on the construction schedule that has been developed. The problem with this is that a construction schedule is informed guesswork at best. Building a construction schedule is like designing a subdivision or lot layout on paper with no elevation contours. It is easy to think through it and design a true masterpiece on paper; then construction begins, and it rains, and people show up late, and shipments are delayed, and inspectors show up late, and it rains some more, and so on the project goes. Because of this, it is best not to use dates, but events. Although a project manager has only an informed guess about when the framing will be completed, he does know that it will be completed, so the point at which the framing inspection should be called in is not June

6, but two days prior to the completion of framing. In this way, the communication plan does not need to be updated every time the project gains or loses a day.

How Should Information Be Sent?

As stated above, communication methods have multiplied greatly in the past few years. This section will address the use—some pros and cons—of various types of communications and offer some recommendations.

Mail The U.S. Postal Service, FedEx, and UPS are still employed in the construction industry, even though there are all the newer forms of communications available. Sending an old-fashioned letter is still considered the best way to give legal notice and ensure that the recipient receives it in a manner that can be documented. Most vendors and subcontractors still send invoices and receive payments through the mail, although some are submitting bills by E-mail and receiving payment through wire transfers. For documents that have legal considerations, mail is still the best way, as it is easy to document receipt through the use of a certified letter.

Cell phone Given the improvements of service, cell phone coverage is almost guaranteed in most areas, which makes this one of the primary means of communications in the construction industry. This is the preferred means of communication when instant feedback is needed. On a side note, a builder can survey which service has the largest stake in the local market to take advantage of in-network calling plans.

Fax Communicating by fax is less common in the industry due to the advances of E-mail, but it is still used by some companies to submit quotes and the like. Traditional fax machines are useful, but project managers might also consider a virtual fax machine. The virtual fax machine routes the faxes to one's E-mail account as documents or image files that can be viewed on laptops and advanced phone devices. E-mail or other documents can also be sent to someone else's traditional fax machine via the virtual fax service. These services can usually be obtained for a low cost each month.

Push-to-talk Push-to-talk is a service that some cell phone providers offer, and it is very popular in the construction industry. It basically turns someone's telephone into a walkie-talkie.

E-mail E-mail has become one of the primary ways by which construction project managers can send out information to project stakeholders. It is a cheap and quick form of communication that provides a paper trail, which the telephone does not offer. If a project manager needs approval for change orders or any other type of modification, and it would be important to have a way of confirming receipt and communication, E-mail is an ideal format. That said, many people only check their E-mail a few times a day; therefore, it may not be the best form of communication if immediate information or response is needed. Often, an E-mail can be sent, and then a phone call placed asking the individual to check her E-mail. This ensures that the recipient receives the information promptly and responds in a way that can be documented.

Private message boards Private message boards (PMB) are the Internet equivalent of a bulletin board. Individuals can determine who has access to which message boards and can post information to the board. Once the information is posted, the members of that board are notified via E-mail or text message that a new message is available. This is good for making general announcements to a specific group of individuals. Some companies develop PMBs for different groups of stakeholders: subcontractors, vendors, customers, project team members, etc. One can also upload documents, which may then be downloaded and referenced. These tools have the ability to communicate in an effective and efficient means. The effectiveness of PMBs, however, will depend on the level of Internet access the various groups have.

Text messages Text messaging is a growing phenomenon among mainly the younger generation. I recently read a statistic that the average teenager sends and receives approximately 2200 text messages per month, which is a startling number. Some project managers are beginning to use text messaging to relay short messages on various items, but this is a technology that should be used sparingly. Because of the limited number of characters with which to compose a message, texting lends itself to misunderstandings, and it is not an overly reliable method.

Social networking Web sites Social networking Web sites such as Facebook, Twitter, and MySpace are some of the fastest growing Web sites in the world. These sites offer the opportunity for people to connect and share messages, pictures, and other types of information. Some builders are seeing the benefits of advertising on such sites, as they offer very focused advertising. But these sites may be leveraged for much more than advertising. It is a great way to communicate with clients. For instance, a project manager can create a Twitter account for a construction project. Vendors, customers, and project team members can subscribe to the account. Whenever there is a project update, the project manager can broadcast the information to either the entire project team or select members. In this way, up-to-date information can be sent to numerous E-mail accounts and cell phones. This method, however, is also currently limited because of the limitations placed on the size of the message. It is my belief that as the next generation comes of age to purchase homes, this will be a very natural and expected means of communication.

Blogs Blogs are the great leveling field of communication. A blog is a Web site where people can write whatever they want and make it available to whomever they want whenever they want. As it relates to a construction project, a blog could be a way to catalogue project events as they take place. Although a blog is a great way to educate the public about various topics and a great way to promote the company, its uses for a construction project are rather limited at best.

Given the numerous options for communication, the project manager needs to develop a means of recording how information should be sent to the various stakeholders. A very helpful tool is a communication matrix. A communication matrix identifies the various stakeholders of the project and states both what their preferred communication method is and how they should receive communications. This is something that can be very easily created in Microsoft Word or Excel. An example is shown in Table 3.3.

A simple matrix such as the one shown in Table 3.3 provides a list of the people and how they like to be contacted. The number indicates the preferred method of communication. The matrix should be accompanied by a project directory that lists telephone numbers, fax numbers, E-mail addresses, or the directory can be included if room permits in the matrix as shown in Table 3.3. By gathering all this information during the planning phase, it will be on hand during the execution

Table 3.3 Communication matrix

Method / stakeholder	Telephone	Fax	E-mail	Other
Gary Lawrence	1. 555-1212		2. gary@	
Tracey Stallings	2. 555-1213		1. tracey@	
Casey Malone	1. 555-1214			
Addison Parsons	1. 555-1215			
Daniel Wheeler			1. daniel@	

phase. It should also be requested that individuals update their information as it changes.

Who Is Responsible for Sending the Information?

The last item to address in the communication plan is the person responsible for sending communications. This person has a vital task and should be up to the challenge. She must be a skilled communicator, both in written and verbal communications. She must not only be able to take a message and relay it, she must have enough knowledge of the construction process to think critically about the information she is communicating.

Earlier this week, in one of my construction projects, a change order was issued by the buyer to change the floor covering on a stairwell from carpet to a hardwood material. The buyer is very cost conscious, so a few different options were offered: unfinished hardwood, prefinished hardwood, or a laminate tread with painted risers. The buyer chose unfinished hardwood for the treads, which would be stained to match other flooring and painted risers. A new employee was tasked with the responsibility for calling in the order for the treads. Unknowingly, the employee ordered treads of the wrong thickness to fit the current stair design. This was a rather simple and insignificant mistake, but it could have been somewhat costly if the incorrect treads had been installed. This simple example shows that the person who told this employee to call and order the treads did not communicate enough information, and the employee did not have the knowledge level to ask the right questions. Everyone just assumed that everyone else knew what they were doing. Assumptions like this can sometimes have insignificant repercussions, but other times they can cause cost increases and time delays resulting in major problems for the construction project.

It is my opinion that many of the problems that arise during the construction project are due to poorly communicated expectations and to misinformation. By developing a solid communication plan that explicitly states how communication will be handled, the project manager can do much to mitigate the negative effects that can occur when communication breaks down.

Risk Management Plan

Risk is a natural part of any undertaking. Driving to work, walking a dog, skydiving, getting married, all carry some amount of risk. Obviously, driving to work is most likely less risky than skydiving, unless one's drive to work is through a war zone. Most people would agree that getting married is much riskier than walking a dog. Every action taken, and even inaction, carries a degree of risk. However, risk in the context of this book is of a specific type.

In this context, risk refers to an unforeseen event that causes a deviation from the project plan. Using this definition, risk could refer to both positive and negative events. For instance, if the cost of a certain material drops during the course of a project, this is a positive risk event. If a subcontractor suddenly goes out of business, this is a negative risk event. It is important that the project manager attempt to plan for all possible risk events, both positive and negative, by creating a risk management plan as a part of the main project plan.

The risk management plan is composed of two major sections. In the first section, the project manager identifies and analyzes risks and plans responses. In the second section, he develops a plan to monitor the various parts of the project so as to identify problems either before they arise or before they have done much harm to the project.

Identify and Analyze Risks

The steps involved in completing the first part of the risk management plan are:

1. Identify possible risk events
2. Assign probability of occurrence and estimate impact
3. Quantify effects of possible risk events
4. Prioritize risk events based on their probability and impact
5. Develop appropriate strategies for managing the risk events (avoiding, transferring, and mitigating)

Identify possible risk events The first step is to identify possible risk events; that is, any occurrence that can cause a deviation from the project plan. This is a very broad definition, so the project manager must use common sense when developing this list. For instance, an invasion by aliens at the project site would most assuredly cause a deviation from the project plan, but this is not a likely occurrence, so it can be disregarded. But vandals coming to steal supplies is a real possibility that should be considered.

The best way to develop a list of possible risk events is through a structured brainstorming session. In this session, team members put forth possible issues that could arise that would alter the project plan. Some of these might include:

- Increased cost of concrete
- Labor shortages (identify specific areas)
- Inclement weather—hurricane, tornado, flood, or snow—damages the structure
- Difficulty acquiring specialty materials

This is just a brief list of the type of events that might be considered. It is helpful to consider risk events according to various categories, such as:

- Structural
- Financial
- Subcontractors
- Suppliers
- Employees
- Clients

Other categories are possible, but the point is that such a method will help the project team think through the project in a thorough and consistent manner.

During this first phase, the project team is not as concerned with discussing the likelihood of the risk events occurring, but rather with simply listing all risk events that they believe could be a possibility. Doing this will take some time, and it might even be helpful to look to some outside sources for information. If a project manager is new to the construction industry, he might consult building inspectors, the local home builder's association, trade magazines, or subcontractors to get a feel for what might realistically be an area of concern. After he identifies the possible risk events, he must consider each risk event more carefully.

Assign probability of occurrence and estimated impact The second step of developing the risk management plan is to consider the likelihood of each risk event's occurring and estimate the impact on the project. The purpose of this is to determine which of the risk events could potentially affect the project's performance. Having a list of possible risk events is not enough; it is also important to have an idea of how likely it is the risk will occur and what impact this could have on the project in terms of time, cost, and quality.

Determining the likelihood of a particular risk event can either be based on hard data or more subjective criteria. Hard data would include past project statistics, as well as any other statistics that might indicate a probability. For instance, assume that one of the potential risk events identified during the first step is the possibility that a vandal could attempt to break into the home and strip the wiring. If someone wanted hard data, she could look back at previous projects and see how frequently this occurred; she could also contact the local police department to learn about the prevalence of this type of crime. A more subjective approach would be to ask the electrical subcontractor how likely he thinks this is. Once the project manager has a feel for the likelihood of a risk event occurring, she then assigns a probability to the event (10 percent, 20 percent, 60 percent, etc.).

Instead of trying to assign individual values to each event, one may want to use a range. This is most easily accomplished through the use of a risk-rating table. An example is shown in Table 3.4. Table 3.4 creates three risk-rating possibilities:

low risk, moderate risk, high risk. Later, however, the probability rating assigned will be used to calculate potential impact. Therefore, instead of providing a range, one might want to have five possible ratings: 0 percent, 25 percent, 50 percent, 75 percent, and 100 percent. This provides a standardized system for assigning risk probabilities. There are incredibly complex models that can be used for such analysis, but for most residential construction projects, the method described here will suffice. In the example, assume that the proposed risk event (vandals stripping wiring from home) is considered a low-risk (10 percent) event.

After assigning the probability, the project manager considers the impact that such a risk event could have on the project. There are two categories of impact: time and money. Some project managers might choose one while others will use both. In this example, the estimated cost of repairing and re-wiring the home, based on an estimate from an electrical subcontractor, is $8000. It would also cost time delays: two days to strip, four days to rewire, one day for inspections. Therefore, it would add seven days to the project schedule. The probability of this risk event occurring is considered low and the impact would be an additional $8,000 in money and seven additional days in time.

Consider another risk event: days lost for rain. By looking at past weather trends and weather forecasts, the project manager can develop an estimate of how weather might affect the project. Let's say the project manager learns that it rains an average of three days per week to such an extent that work on the project will have to stop. This type of weather reliably occurs for the whole month of November, giving it a risk rating of 80 percent. If the month has 30 total days, 8 are nonworking days (weekends), and 22 are working days, approximately 26 and 74 percent, respectively. Given that 3 out of every 7 days are nonworking days due to rain during this month, there are approximately 13 total days that are nonworking. Multiplying the 13 days by the 76 percent (percentage of working days during 30-day period), gives an estimate of approximately 10 days lost during that particular month due to rain. The risk event impact then is the loss of approximately 10 days due to inclement weather. This is a much more helpful expression of impact than simply "days lost for rain."

Table 3.4 Risk rating table

	Low	Moderate	High
Likelihood	0–30%	31–60%	61–100%

The project manager must go through each risk event and determine its probability of occurring and its impact on the project. If the risk event has a high probability (heavy rain in November) and a high impact in cost or time delays, the project manager should take the time to analyze the factors and attempt to arrive at some informed estimates. If the risk event has a low probability (alien invasion), the project manager need not take time to analyze that particular risk event. The project manager must use a certain amount of discretion in determining which risk events to analyze. Events with high likelihoods and low impacts may or may not need to be analyzed. Or events with a potentially high impact, but a low likelihood may or may not need further analysis. The project manager will decide based on her judgment.

Quantify effects of possible risk events   The next step is to quantify the effects of the various risk events. This is a relatively simple step that attempts to assign value based on the likelihood of a risk event's occurring. In the first example given previously (vandals), the project manager determined that the probability of the home's electrical wiring being vandalized was low, or 10 percent. The cost of the rework was estimated at $8000 and the loss of seven working days. To quantify the effects of this risk event occurring, the project manager must simply multiply the probability by the impact: $8000 multiplied by a 10 percent chance equals a cost of $800. Therefore, the cost impact makes this a relatively low-level risk event.

In the second example (loss of days due to rain), the project manager determined there was an 80 percent likelihood of 10 working days of the month having too much rain to work. Therefore, it is likely that 8 days will be lost to inclement weather. This risk event could also be quantified as a dollar amount by assigning a value to each day's work based on various contributing factors that are unique to each project. Contributing factors could be penalties contained within the contract or the cost of rented equipment. For sake of the example, assume that each day is worth $1000. Therefore, the project manager could assign an $8000 cost to this risk event.

By quantifying the risk events, the project manager ascertains how much attention should be given to each risk event. Those risk events that have a high probability and a high level of impact should obviously be given greater consideration than those risk events that have a low likelihood and a low level of impact. The only way to know which is which is to perform this type of analysis. A very helpful tool for preparing this portion of the project plan is a spreadsheet that outlines the information related to each possible risk event. The format for such a chart is shown in Table 3.5.

This type of chart brings everything together in one place where it can easily be referenced and analyzed. It's also easy to update as the project progresses or new information becomes available. If a project manager uses a software package

Table 3.5 Risk event 01

Risk event	Likelihood	Impact (days or $)	Quantified effect
Rain delays	80%	10 Days	8 Days
Wiring vandals	10%	$8000	$800
Alien invasion	0%	$200,000	$0

such as Microsoft Excel to create this type of spreadsheet, he can easily calculate impacts. This also makes it easier to check various scenarios by simply plugging in the numbers from the various scenarios.

Prioritize risks based on their probability and impact The next step is to prioritize risk events based on their probability and impact. Some risks are greater than others and need to be treated with a higher priority than others. If the information has been entered into a risk event spreadsheet, it is relatively easy to arrange them once they have been prioritized. Risk events that have a high probability of occurring and a high impact are given priority.

There are a couple of ways that the chart can be arranged or coded to show which events have the highest priority. One method is to first categorize the risk events by type: structural, electrical, plumbing, etc. Then order the associated risk events within each category according to priority. This helps in locating the right information associated with each area of the project. A new column can be added to the spreadsheet that shows the priority level given to each risk event, which would appear as shown in Table 3.6.

Table 3.6 Risk event 02

Risk event	Likelihood	Impact (days or $)	Quantified effect	Priority
Rain delays	80%	10 Days	8 Days	High
Wiring vandals	10%	$8000	$800	Low
Alien invasion	0%	$200,000	$0	Zip

The convention used to show priority in Table 3.6 is titles, such as high, low, or moderate; but one could also use a numbering system. A numbering system might be advantageous if a project manager is creating this type of table in a spreadsheet program that has a sorting function. By assigning a number in a spreadsheet application, she can sort the table by priority, likelihood, impact, or any combination of these. The benefit is that she will be able to query the information a few different ways in order to make certain that no important information is overlooked.

Plan Responses to Risks

Now that each risk event has been prioritized, it is time to develop strategies for mitigating those risk events that pose a reasonable threat to the success of the project. In this phase, the project manager develops a risk response plan for each of the possible risk events. There is a variety of strategies that she might employ, depending on the type of risk being considered: positive or negative. A negative deviation harms the project in some way. A positive deviation benefits the project, and these types of events should be considered as well. Strategies must be developed to work with both types of risk events. Typically, positive risk events are referred to as opportunities. The primary strategies for dealing with a positive risk or opportunity are enhancement and exploitation.

Enhancement The enhancement technique seeks to increase the chance that the opportunity will present itself. The technique may be used in a variety of ways. The first step is to recognize that a certain event could transpire. The possible event is then analyzed with the goal of increasing the probability that the event will take place, resulting in a positive impact on the project plan. For instance, if the project manager were to learn that future zoning changes would positively affect a project, then he could begin lobbying for those zoning changes to move forward as quickly as possible. If he were to discover that lumber or concrete prices were going to drop shortly, then he might delay an order temporarily in order to secure the better pricing. These are only two examples, but the possibilities are endless.

Exploitation The second technique for managing opportunities is exploitation. This technique is employed when the project manager wants to make certain that the full benefits of an opportunity are realized. The enhancement technique increases the probability of the opportunity becoming available; the exploitation technique is used to make certain that full advantage is taken. For instance, if the contractor negotiates a special rate with a subcontractor for one particular job, she might try to secure the same pricing for other jobs, thus fully exploiting the situation. The term *exploitation* has overtones of unethical behavior, but this is not what is meant. The question is not how one can take full advantage of a person, but of a business opportunity, assuming that the company acts in an ethical manner.

Typically, there are many more negative risk events to contend with than there are positive risk events. A project's success will greatly depend on how well the project manager manages the variety of situations that arise during construction which threaten some aspect of the project. A project manager who is constantly reacting without careful consideration will fare much more poorly than the one who anticipates problems and creates response plans. In creating these response plans, the project manager has three primary strategies for dealing with negative risk events, avoidance, transference, and mitigation.

Before each of these techniques is considered in turn, it is important to note that they are not mutually exclusive. In many instances, the project manager will employ some combination of all three in managing various risk events. So when analyzing each risk event from the risk register, the project manager will consider how each technique might work to minimize the impact or lower the probability of each risk event.

Avoidance The first technique for managing negative risk events is avoidance. Some risks are unavoidable; they must be faced if the job is going to be done. This is the very nature of risks and projects. However, some risks can be avoided entirely. The exact shape that this technique takes will vary depending on the type of project or the type of task that is being considered. This is the tactic that some builders have taken from the very beginning when considering a project during the initiation phase. They may meet with the prospective client, review the plans, consider the job site, then decide that the project is so fraught with risks—the plan is poorly conceived, the budget is unrealistic, the clients seem unreasonable, the project seems unprofitable, the project is beyond the expertise of the contractor, and so on—that they refuse to take on the project. It is a wise contractor who knows when not to take a job.

Assuming the project has been accepted, the project manager is seeking to learn if there are any risks that have been identified that can be avoided altogether. One risk might be that the buyers want an in-ground pool in the back yard, but the project manager knows that the pool will be close to the water table, so he convinces the buyers to install an above-ground pool. In this instance, he has avoided digging the hole and risking hitting the water table. Avoidance is probably the least common technique used, because the project manager cannot simply choose what work to do or not to do, so he will make much more use of the next two techniques: transference and mitigation.

Transference Most often a particular risk cannot be avoided entirely, but the burden of the risk can be transferred to another party, who must then shoulder the risk. This is a type of avoidance, but it is referred to as *transference*. The clearest example of this is insurance. In the risk table, weather, vandals, job site accidents, and the like will most certainly be listed as potential risk events. These risks cannot be entirely avoided, so they must be managed in some manner. By purchasing insurance for a relatively small cost, the risk of the adverse events taking place is not necessarily reduced, but the impact that the event could have on the cost of the project is significantly reduced, as the insurer assumes that risk.

Transference may also be employed through the use of subcontractors. If the plan calls for a particularly difficult bit of work that the builder feels unable to perform to the quality standard demanded, she might hire a specialized subcontractor to perform the work, thus transferring the risk associated with the work to the subcontractor. The builder still has some risk even in this situation, as she takes on the risk of choosing the right subcontractor, but the overall risk for this particular task should be greatly reduced.

Mitigation The last technique to consider is the one that is probably most commonly employed: mitigation. If a risk cannot be avoided, if it cannot be transferred onto a third party and must be met head on, as it were, by the builder, the builder should take steps to at least mitigate the effects or the probability of the risk. Once again, the exact form this mitigation will take is quite diverse. The project manager may even choose to use a couple of mitigation techniques on a single risk event. Consider some simple examples:

- The roof is particularly steep and treacherous, but the roofing must be done, so the roofers will tie-off with harnesses and straps to mitigate the risk of falling off the roof.
- The building inspector is very particular, so the project manager takes special care when dealing with him, asking detailed questions to make certain that the work is completed to satisfy the inspector.
- There is some concern that a back wall will lack the strength over the long run to endure the pressure on it; the wall is reinforced with steel or concrete or a higher grade of lumber to ensure long term endurance.

These three examples demonstrate how one can mitigate risks through simple steps. Recall the previous example of vandals. If vandalism is a major concern for a particular job site, the project manager might purchase insurance to cover any losses resulting from vandalism, which is the transference technique. The project manager might also want to mitigate the probability of this event even occurring. In this instance, the insurance company would be able to pay for the repairs necessary, but the insurance company cannot make up for the loss of time. Now it may be true that the contract has extension clauses in such instances, but the project manager would like to finish the project on schedule. So she might choose to also take steps to mitigate the probability that the vandalism even occurs. She can do this through placing a night security guard at the job site or hiring an offsite security firm to make sweeps every so often. Depending on the project size, this might not be a cost-effective measure. However, it might be cost effective to place security cameras throughout the job site and large signage announcing their presence. This will deter any vandals and thus mitigate the probability of the vandalism occurring.

Monitor Risks

So far, risk events have been identified and analyzed, and responses have been planned. This comprises the first half of the risk management plan. In the second section, a plan is developed to monitor the various parts of the project in an attempt to identify problems either before they arise or before they have done significant harm to the project. In order to complete the second section of the risk management plan, the project manager will need the first half that has just been completed, and he will need the WBS, the project schedule, and the human resource plan.

Recall that the purpose of a risk management plan is to determine how risks will be managed while the actual work of the project is progressing. The best way to do this is to identify when a risk event is most likely to occur during the life of the project. For instance, if one of the risk events involves a certain inspection, then it is important to monitor construction more closely up to that inspection. After the inspection has passed, then the time window of the risk event has passed. Not all risks involve a closed time window. Some—such as weather, subcontractor issues, and the like—may be possibilities throughout the life of the project. Nonetheless, it is important to determine when a risk event should be monitored more closely. This requires determining when the risk is likely to occur, who will be responsible for monitoring the risk event, whom to report the results of the monitoring to, who can authorize the response plan, and who has the responsibility for monitoring the implementation of the response plan.

If one is dealing with a relatively small project in which the project team is small, this may simply be the responsibility of the project manager or of one other person. The written plan only needs to be as detailed as is necessary to ensure that the monitoring is handled properly. If it is a larger, more complex project, the monitoring may be assigned to different individuals according to his knowledge and abilities. For instance, someone who has virtually no knowledge of electrical wiring should not be the person monitoring a risk event that is specific to the electrical plan of the home. This second part of the risk management plan could be displayed in another spreadsheet like the one used in the first half of the risk management plan.

Purchasing and Contract Administration Plan

A major portion of the construction project is purchasing materials and services and administering the contracts that regulate those purchases and other project work. It is a behind-the-scenes portion of the project that can be a strong contributor to project success or project failure. Purchasing materials at the wrong time, not understanding what exactly is being purchased, hiring the wrong subcontractor, or not effectively managing the contracts with subcontractors and suppliers can lead to project delays, cost overruns, and quality control issues. The project manager must ensure that a strong plan is developed and that the plan is followed by a qualified individual. The actual plan includes purchase, subcontractor, and contract management.

Purchase Management

The first part of the plan deals with purchase management—purchasing of materials and building supplies. Earlier during the planning phase, the project manager estimated the materials needed to perform the work of the project based on the construction plans and specifications. He might or might not have gathered these quotes from the companies that will actually provide the materials used to construct the building. The quotes were a means of preparing the project budget. Now the time comes to actually plan the purchase of these materials. This portion of the project plan is concerned with:

- What materials are needed?
- When are they needed?
- What is the budget for the materials?
- Who authorizes the purchase of the materials?
- Who inspects what has been received?
- Are there any special instructions?
- Who is responsible for managing purchases?

The following project documents provide the information necessary to know what materials are needed:

- WBS
- Construction plans and specifications
- Cost estimates

From these documents, the project manager should be able to identify the majority of the materials needed to construct the home. He should realize, however, that these lists are not comprehensive, for a couple of reasons. First, much of the material necessary will be provided by the subcontractors themselves. Plumbers and electricians, along with a host of other subcontractors, typically include materials as part of their quote for the work. If the contract with the client requires specific materials for the bathrooms or other areas of the home, the project manager must make certain that the appropriate subcontractor includes those specific materials as part of the quote. The second reason that not all the materials will be listed is that there are some materials that cannot be estimated until the work begins. For instance, before the rough-in inspection has been completed, the home must be fire-caulked. How many tubes does this take? It depends on how tight the joints are, how many points of entry there are to certain areas, and who is doing it. There should be a line item in the project budget for general, incidental items. But from all the previous work done for the project plan, it should be fairly easy to pull together a materials list. If the contractor relies heavily on subcontractors, the list may not be extensive, but it will typically include the following:

- Concrete for footing, foundation walls, basement slab, driveway, any patios
- Block for crawl space or basement wall or main wall of home, depending on the design plans
- Framing, roofing, and window package
- Insulation materials
- Drywall supplies
- Paint
- Flooring materials
- Cabinetry
- Light fixtures
- Appliances

This is by no means a comprehensive list, but it shows some of the major expense items that are typically paid for directly by the builder. Most of the other materials are provided by the subcontractor as part of their services.

A second part of purchase management involves determining when the materials should be delivered. Some suppliers will allow a builder to purchase the materials at the quoted rate and then hold them for pick up or delivery when they are needed on the job site. This is the preferred method, as the builder does not have to move the materials to a warehouse and store them. Storing materials at the job site always increases the risk of having the materials either vandalized or stolen, thus the project manager can mitigate this risk event by storing as little as possible on the job site. Knowing when to order materials requires knowing the lead time necessary for delivery or pick up and knowing when the materials will be needed on the job site. The project schedule will act as a general guide, but the project schedule will continue to be updated as construction progresses. Therefore the purchasing agent must closely monitor the progress of the project to know when to place the order for materials.

Third, the project plan should specify what the budget is for the materials and who is responsible for authorizing purchase orders for the materials. The budget information should be included as part of the primary project budget, but it is helpful to insert it into this portion of the project plan as well so that it is readily accessible. The project manager will determine who can authorize purchases. This person should not take the job lightly. A mistake could cost thousands of dollars and weeks in construction delays.

Another important consideration is order confirmation. Someone must make certain that the materials delivered match the materials ordered. If a certain species of wood of a certain grade is specified in the project plan and an inferior grade is delivered, then major problems could arise if the mistake is not caught. If someone is assigned to inspect and confirm all deliveries, the risk of the wrong materials being accepted is greatly reduced.

One person should be ultimately responsible for overseeing all purchases. This may be the project manager in smaller projects. By having one person oversee all aspects, the opportunity for deviation from the project plan is greatly reduced.

Subcontractor Management

The second part of the purchasing and contract administration plan involves managing subcontractors. How often the general contractor deals with subcontractors will determine how much information goes into this section of the project plan. The best way to organize the subcontractor management plan is by providing a basic flow for how subcontractors will be managed, then ensuring that the plan is followed. A basic flow may appear as follows:

- Determine which subcontractors have necessary expertise
- Request quotes from approved subcontractors
- Select subcontractor based on quality and value of work proposed
- Enter into contractual agreement with subcontractor
- Coordinate scheduling with subcontractor

- Subcontractor performs work at appropriate time
- Inspect and approve subcontractor's work
- Approve payment per terms of contract

The flow is fairly straightforward. The key is making certain that someone who can make sure that the process is properly followed is managing the process. Again, on smaller projects, this will most likely be the project manager, but it could be a project team member who possesses the knowledge and skill necessary to manage the work.

In selecting subcontractors, there are a number of issues that should be considered: quality of work, cost, expertise, etc. Legal guidelines must also be considered. In hiring and working with subcontractors, the general contractor must make certain that he obeys all applicable hiring and antidiscriminatory laws, on both the federal and state levels. Dealing with these issues is beyond the scope of this book, but the builder must make certain that all laws are followed to avoid any ethical or legal misconduct.

Contract Management

The next section of the purchasing and contract administration plan is contract management. This section involves both vendors and subcontractors, because contracts are employed with both groups. Contract administration involves the creation, management, and closing of contracts. Most of the time, vendors who provide materials write the contracts that are followed, which is typically outlined in their return policies. The project manager should make certain that she is aware of how the vendor will handle complaints and returns. Typically, there are standard return policies across the industry based on the type of material.

The more pressing contractual issues involve the client and subcontractors. The contract with the client, which most likely has already been signed when someone reaches this point of the project planning phase, must be managed just like any other contract. There are requirements on the part of the builder and the client which the builder must manage in order to ensure that the project is in consistent compliance with the project's contract. If the builder is out of compliance, it is important that he realizes it as soon as possible and takes steps necessary to become compliant. If the client is out of compliance, the builder must address this issue with the client. It may be as simple as reminding the client to ensure that certain choices are made by a certain date and the like, or it might lead to a dispute, which halts construction and ends after months or years in arbitration or court. By consistently managing the contract with the client, the project manager can ensure that all parties realize that each person will be held accountable to the written agreement. If the client or the builder verbally agrees to any variation from the original contract, then it should be written and amended to the project's contract.

The contract with the subcontractor is typically shorter and less involved than the contract with the client, because it only deals with one or two aspects of the project. The goal is to protect both the builder and the subcontractor. The contract should name the parties, the price, the work, the time frame, how disputes will be handled, and who is liable for what. If the builder changes the work, the contract must be amended and all necessary areas addressed, such as the price. If the subcontractor attempts to change the price after the work is completed, the builder has the benefit of the contract to show the agreed-upon amount. Sometimes a subcontractor will argue that she was unaware of the fact that she would have to do this or that. If she goes ahead and does it without discussing it with the builder, the builder cannot be held responsible for paying for the extra work. It is incumbent upon the subcontractor to accurately estimate and quote the work.

One of the main areas that will be addressed in the contract with the subcontractor is liability: Who is responsible for what if something goes wrong on the project. Through the contract, the builder should attempt to transfer as much risk from herself and onto the subcontractor, thus mitigating her exposure. The subcontractor, however, should attempt to take on as little risk as possible, thus reducing his exposure. The builder should require and verify that the subcontractor has the necessary types of insurance and make it a part of the contract as a means of reducing the builder's liability.

Managing the contract with the subcontractor is not much different from managing it with the client; it is just on a much smaller scale. The person assigned to managing contracts should verify that the contract is being adhered to by both parties. Whenever a breach occurs, the project manager will have to address the issue and seek to resolve it in a manner that has the least negative impact on the project.

The next section of managing contracts is applicable to both the client and subcontractors: change management. Change orders are typically a source of frustration for most builders. This does not necessarily have to be the case. Changes that are managed well will delay the project some, increase the cost to the client, and cause some frustration, but they do not have to be nightmares. Clients and subcontractors should be discouraged from making changes to the project plan once the contract is agreed upon, but once they are definitely going to make a change, the sooner it is known the better.

Changes to the project plan are costly. They cost time, money, resources, and sometimes quality. A simple change creates a ripple effect that can produce a great wave by project end. Most clients and many builders fail to consider the true cost of a change to the project plan. To the client, it may seem to be nothing more than moving one wall over 12 inches, but the builder sees the following:

- Subcontractors to revisit job site
  - Flooring
  - Framing
  - Drywall
  - Electrical
  - Plumbing
  - HVAC
  - Painter
  - Trim carpenter
  - Building inspector
- Modifications to project plan
  - Scope
  - Cost
  - Schedule
  - Quality inspections
  - Contract management—negotiate amendments to subcontractor contracts and client contracts
  - Risk management
  - Change management
  - Purchasing

With this one small change almost every aspect of the project is affected and shifted to accommodate the change request. Because of how much a change order can affect a project, the cost is typically three or four times what it would have been if the plan had originally included the change. This can lead to conflict with clients, subcontractors, and confusion and frustration within the project team. The only consolation is the fact that the builder has included some profit in the change to compensate for all the frustration.

Changes can be avoided by one of two methods. First, the builder can simply state within the contract that no change order will be accepted. This is a practice employed by some large-scale builders who work with small margins and cannot afford the time or the cost to deal with change orders. The second method is that the builder can spend much time on the front side planning the project in great detail in an attempt to make certain that the client has made an informed decision about the various options available. The builder should also take the time to explain why change orders are so costly. If the client realizes the delays and costs that can result from change orders, she is less likely to request one. If the client does request a change order, the builder needs to have a plan for how it will be addressed.

Creating a change order is almost like creating a construction miniproject. It requires the same phases that the primary project does: initiation, planning, execution, controlling, closing. During the initiation phase, the client presents the change order and the builder provides a cost estimate and an initial report on how the change order will affect the project. If the client decides to move forward, the project manager enters the planning phase. He develops a plan for performing the work and makes the necessary changes to the project plan. Third, he executes the change order by performing the actual work according to the modified project plan. As the work progresses, it is monitored to make certain that it conforms to the project plan. After the work is completed according to the specifications, and the work has been verified, it is ready to be inspected by the interested parties. Once the client approves that the work was done according to the agreement, the change order is closed. An individual should be assigned to oversee the entire process to make certain that the implementation of the change order is done properly. Along the way, the project manager should update the project plan to reflect the fact that the change order is being implemented. This topic shall be dealt with much more extensively in Chapter 5.

The last portion of contract management deals with closing contracts. Closing contracts refers to taking those steps necessary to ensure that all parties have met their obligations according to the contract. This takes place on small-scale and larger-scale aspects of the project. For instance, when materials are delivered to the job site, they should be inspected to make certain that the actual materials delivered match the materials requested. Once this is confirmed, it should be noted that the materials received were in fact what was ordered. This will, in effect, close the contract, given that no defects arise. After a subcontractor performs the work on the project and the work is inspected, she is to be paid if the work is in compliance with the terms of the project, and the contract is to be closed. This is to say that the parties have fulfilled their duties as outlined in the contract. On a much larger scale, the construction contract with the client will be closed on completion as well. This typically takes place after the certificate of occupancy is received from the local building inspection office, and the final walk-through is completed. Sometimes this takes place at a closing table at an attorney's office, or the client can simply sign off on a statement that the builder has completed the project according to the specifications detailed in the contract. Regardless of the method, it is important that the builder receives written confirmation from the buyer that the builder has fulfilled his contractual obligations, just as the buyer should want the same from the builder.

There are many problems that can arise when managing and closing contracts. Builders are typically very familiar with contracts and their terms, because they deal with them on daily basis. Buyers are typically much less familiar with these matters. The opportunities for confusion and misunderstandings are numerous. If the builder believes the client or any other party to be in violation of the contract, she should discuss the matter in a professional manner. If the matter cannot be resolved through a simple discussion, the builder should seek legal counsel. Clear communication regarding the terms of the contact so that everyone knows what to expect is an important means of avoiding conflict. Most contract disputes arise from misunderstandings that could have been avoided if more time had been taken to discuss the contract.

PROJECT DOCUMENTATION AND PROJECT BASELINES

This section of the project plan is primarily a reference section. First, it contains the documentation that guides the project, such as:

1. Project contract
2. Construction specifications—blueprints and building plans
3. Site survey
4. Building permit, insurance policy, septic tank approval and installation guidelines, sedimentation control plan, watershed control plan, and other necessary documentation
5. Subcontractor and vendor work contracts

The purpose of including this information is to make certain that it is easily accessible throughout the life of the project. Too many builders do a poor job of keeping up with these vital documents which guide the construction project. It is often a good idea to include copies in the project folder and keep the originals in a separate location in case something is misplaced. As construction progresses, these documents can be referenced during inspections and discussions with subcontractors to make certain that they understand how they are to perform their work.

This section also contains the baselines pertinent to most construction projects, such as the cost and schedule baseline. These baselines provide a basis for comparison as the project progresses. By comparing actual results with planned results, the project manager can determine whether the project is on schedule or within the cost guidelines. Failing to compare actual results with planned results can lead to unexpected delays or cost overruns, which might have been avoided if the project manager had taken the time to analyze the project's progress. The method used to do this was discussed briefly in the cost control section of the project plan, but more details will be provided in the chapter discussing the controlling phase of the construction lifecycle.

TRANSITIONING TO THE EXECUTION PHASE

In this book, more time is spent on the planning phase of the project than the other portions of the project lifecycle. This is not because the other areas of the project lifecycle are unimportant. It is because if more time is invested in the planning phase of the project, less work will be required down the line. This may seem incorrect, as nothing gets physically built during the planning phase, so most contractors do not see the point in spending so much time doing it; they would rather get to the execution phase. If the time is invested during the planning phase, the execution phase is simply a matter of executing the plan while making the necessary adjustments that will most likely be required. As the project manager becomes a better planner, the adjustment will decrease. By spending the time necessary to thoroughly plan a project, the project manager can dive into the execution phase knowing that he is prepared for the work at hand.

REFERENCES

Leach, Lawrence P. 2005. *Critical Chain Project Management,* 2nd ed. Norwood, MA: Artech House.

4

EXECUTING THE
CONSTRUCTION PROJECT

So far much work has been invested in the construction project. The project has passed through the initiation phase and most, if not all, of the planning phase. The project team has been researching different aspects of the project, analyzing potential problems, developing solutions, and is now ready to see the actual work of the project begin. Everyone is ready to see that proverbial first shovel go into the ground.

In some ways, the execution phase is the easiest and the most difficult phase of the project. It is easy in that now the project manager must simply execute the well-developed plan. But it is the most difficult because she must manage the work. Anyone who has ever started a small or a large business venture knows this experience. The lemonade stand proprietor, the real estate agent, and the contractor all face the same challenges: They must take the plan they developed on paper or in their mind and make it a reality. Almost any business plan can be made to look great, but there are two questions: Is it realistic, and can the person work the plan? The entire planning phase has been spent making certain that the plan is realistic. The work has been planned well; now the time comes to work the plan. This requires both discipline and flexibility. If a project manager lacks the discipline to follow the plan, he will most certainly fall short in some way. Likewise, if he is unwilling to adapt the plan to real world conditions, he will be disappointed again and again.

This chapter focuses on guiding the project manager through the project plan, showing him how to execute the plan that has been developed. It does not focus on specific building practices or the like, but on how to execute a project successfully: the personal skills that are necessary, as well as the process one uses to actually move through the project plan. First, the characteristics needed for successful project execution are discussed.

CHARACTERISTICS NEEDED FOR SUCCESSFUL EXECUTION

Executing a construction project plan is more complex than the plan may reveal. The project plan sets out what is going to be built, how it will be built, who will build it, when it will be built, and a host of other matters. However, there is an unspoken expectation that is built into a project plan—that a competent project manager executes the project. Once the project begins, a number of issues and competing interests will arise. Stakeholders will push for their agendas, subcontractors will seek extensions or increased pay, employees will get sick and miss work, building inspectors will be difficult, customers will change their minds, and about everything that could go wrong will go wrong. The project manager has to hold the project together and manage all these competing interests and events without losing heart or mind.

A project manager's ability to succeed at the project greatly depends on his ability to master a few skills and personal characteristics. This is not to say that someone cannot finish projects without these characteristics, but it is unlikely he can do so with a high degree of success. If success is merely completing the project on time, within budget, and to the quality standard requested, then a project manager can lack these characteristics and complete projects successfully. But this is a low standard. What if those three things are accomplished, but the client cannot stand the project manager, subcontractors do not want to work with him again, vendors believe he's undependable, and other project team members hope to be assigned to work with a different project manager? Is the project still considered a success? Did the project manager who bulldozed over everyone with a *my way or the highway* mentality really achieve project success? I contend not. This project manager failed miserably, not only by hurting his reputation, but also by damaging the reputation of the company and of the industry. If he had worked to develop the following five characteristics, it is much more likely that the project could be considered a true success. The highly successful project manager must be:

- Disciplined
- Highly aware
- Service oriented
- A problem solver
- An effective communicator

These five characteristics represent the softer side of the skills that a project manager should possess. In the following sections, each characteristic is dealt with individually, and then more direct application is made to executing the project plan.

Disciplined

Personal and professional discipline is something of a lost art, which, appropriately leveraged, can yield great benefits for the project manager during the execution phase of the project. Once the project begins and the pressure begins to mount from various stakeholders, the project manager must exercise the discipline necessary to make certain that the project progresses according to the plan with as few deviations as possible. If the project manager plans the work, but fails to work the plan, then she has not only wasted hours of planning, but dramatically increased the possibility of failure.

The project manager should exercise discipline in a number of areas. First, the project manager must be disciplined in the approach taken to executing the project plan. The plan has been created with much thought, and its purpose is to ensure success. If the project manager shows a willingness to deviate from the project plan without sufficient reason, then others will do the same. This creates confusion and uncertainty. The project manager appears erratic and unpredictable, or even untrustworthy. By adhering to the plan, the project manager ensures that the work is done according to the agreed upon plans and specifications.

Second, the project manager must be disciplined in the approach taken to working with project team members, subcontractors, vendors, and other stakeholders. This means holding people accountable for their performance on the job. A project manager who fails to hold vendors, subcontractors, employees and other stakeholders accountable for their performance will most likely see the quality, the budget, or the time schedule begin to slip. He must make certain that people do what they are supposed to do, when they say they will do it, and how they have agreed to do it.

Highly Aware

The project manager must also exercise a high level of awareness to the ebb and flow of the project. This awareness refers to both the progression of work and the interaction of stakeholders. First, in regards to the progression of work, the project manager must be sensitive to problems or areas of concern. What might begin as a small issue might grow into a major problem. If a worker is doing work poorly on a relatively minor part of the home, what should lead the project manager to believe that the employee will be more competent on the major areas? The project manager must also evidence awareness of how shifts within the project will affect the budget, time schedule, and quality of the project. This requires her to be intimately knowledgeable of every area of the project as work progresses.

Second, the project manager must show a high level of personal awareness of how various stakeholders interact in relation to the project. Conflict between stakeholders is inevitable. Vendors will deliver the wrong materials, subcontractors will show up late, and clients will be upset over something or another. This is the nature of construction. The project manager must manage all these competing interests and tensions without contributing to the problem as the project progresses. This requires listening to what people say and how they say it. If the project manager picks up on some information that could directly or indirectly affect the success of the project, then she must address it. Failure to do so could cause problems later on.

Service Oriented

Unfortunately, to some project managers, a construction project is merely an opportunity for them to exercise their dominion and control over the peons below them. This type of project manager requires a wide berth and leaves a path of destruction in his wake. He confuses a brash personality and no-nonsense approach with leadership. He is still the bully that he was in fifth grade; he just has a more sophisticated means of extracting everyone's proverbial lunch money. The industry, unfortunately, has too many of this type of person.

In a service-oriented approach to leadership, the project manager does not see himself as a tyrant to rule his subjects, but as a facilitator who seeks to use appropriate means to accomplish the project. Instead of looking for what everyone can do for him in order to complete the project, he seeks opportunities to help others accomplish their work successfully so that the entire project is a success.

Service-oriented leadership measures success by not just achievement of the goal, but the means used to get there. This type of project manager seeks opportunities to help vendors, subcontractors, employees, and clients succeed in their individual portions of the project. The project manager asks what he can do to make everyone's job easier and less frustrating. This is most certainly, however, not a call for lowering expectations for performance. Even in this approach, the project manager must keep the performance bar high, calling all of the workers to perform in accordance to their obligations. The difference is that the service-oriented project manager strives to make certain that all involved have every opportunity to perform their parts of the project well.

A Problem Solver

The ability to solve problems is one of the most important characteristics that a project manager needs to succeed. The project manager will face a multitude of challenges and problems as the project progresses. Whether she is able to solve these problems will, in part, determine whether the project is a success. These problems will range from the technical to the interpersonal. A subcontractor will discover that he is not able to do what he originally thought he could do because of this, that, or the other. Two employees will become enraged at one another in some childish turf war. The problems will range from relatively minor to somewhat major, and the project manager will be expected to find a solution.

Being a good problem solver requires two things: (1) knowing what questions to ask and (2) knowing where to get solutions. Knowing what questions to ask is the most important part of solving a problem. If the project manager does not know the right questions to ask, then she most likely will never discover the real problem. She will become lost in treating symptoms without ever dealing with the root cause. Questions reveal more questions, which reveal more questions, which eventually lead to root issues. If the project is continually growing behind schedule, then the project manager should begin asking why this is so. If at first she thinks the subcontractors are the primary issue, then she should seek to discover why. This requires asking all the necessary questions of all the right people. It might be because an incompetent individual is doing the scheduling, or it might be that the subcontractors are not honoring their agreements; the manager can only learn the answer by asking questions.

After asking the questions, the project manager must develop a solution. Sometimes, this is as simple as making a decision. Other times, however, it requires letting people go or shifting their responsibilities. Sometimes, it will require hiring a specialist such as an engineer. The project manager should not expect to be able to fix every problem by himself, but he should know where to look to find a solution. He might find it by asking for feedback from the project team, or the subcontractors, or a specialist. But he must be willing to ask for help. After hearing sound advice, he has the responsibility to make decisions and take actions that will solve the problem.

An Effective Communicator

This characteristic is probably the most important one of the five. Effective communication skills are critical to the success of the execution phase of the project. They are just as necessary for the initiation phase and the planning phase. However, the effect of communication is different in the execution phase. Once the message goes out, the work is started, and the results are seen. If the project manager has not effectively communicated the instructions, the work must be redone; there is no time to modify or reflect on the plan. As a matter of fact, the execution phase could almost be called the communication phase, as the project manager's primary job is to communicate directives based on the project plan.

Communication is directed at a number of groups: employees, project team members, vendors, subcontractors, the client, building inspectors, and other project stakeholders. Each group requires different levels of communication at different times and through different means. From the project plan's communication section, the project manager knows who needs what information and when, but she must still make certain that the information is received and understood clearly. This requires the project manager to not make assumptions and to risk repeating herself and asking others to repeat back to her the instructions to make certain they understand.

The project manager needs to know his personal communication strengths and weaknesses. Some people are very effective communicators in person, but they have a difficult time putting their instructions into writing. Others are excellent writers, but have difficulty when communicating face-to-face. The project manager cannot afford to have these handicaps. He must be able to communicate through any means necessary in an effective manner. If he cannot, then he must find someone who can and get her to review all communications. Instructions that are misunderstood can cost the company more than just money; it can cost the company its reputation, which can result in the loss of future projects.

The project manager's ability to communicate clearly and effectively is key, but there is more. The project manager should also foster effective and open communication among the stakeholders of the project. There are some environments in which no one speaks of problems or concerns. No one wants to talk about the elephant in the room, regardless of what it is: structural issues, contractual issues, interpersonal issues, or whatever. Everyone turns a blind eye and moves along in their roles as though there is no problem—because the project manager will not address it and everyone follows along. This type of environment is like a cancer that eats away at the project team and the project until it is finally killed.

The project manager, however, should avoid this type of environment at all costs. Problems caught early can typically be corrected, whereas those that have time to grow might snowball into catastrophes. By fostering an open environment in which communication is encouraged, the project manager can help ensure that the right information gets to the right people at the right time.

PROJECT EXECUTION

Executing the Work Breakdown Structure

The project manager has initiated the project and prepared the plan; she believes that she possesses those characteristics that will contribute to the successful completion of the project. Now what? The time has come to begin the actual work of the project. The next section discusses each step of the WBS that was prepared during the planning phase of the project. The remainder of the chapter provides a practical guide for moving through the WBS during the construction process.

Note that at the conclusion of each phase, the final step is an inspection by either the project manager or building inspector. Particular points for attention are discussed, but there are two activities common to all inspection phases: once the work has been approved, the project manager issues an approval for payment—either partial or full—to the subcontractor. Then she updates the schedule, taking into account any delays or accelerations that have occurred, being sure to inform any others who need to know about modifications. As these steps are part of every inspection phase, they will not be repeated in the text, but they must be performed by the project manager.

The WBS would appear in outline form as follows:

1.0—Construction project A
1.1—Permitting and site prep
    1.1.1—Prepare a site plan
    1.1.2—Acquire permits
    1.1.3—Prepare site
    1.1.4—Phase inspection
1.2—Foundation, framing, rough-in
    1.2.1—Foundation work
    1.2.2—Framing
    1.2.3—Rough-ins (HVAC, plumbing, electrical, other)
    1.2.4—Exterior work (siding, trim, brick/rock, etc.)
    1.2.5—Phase inspection
1.3—Interior finishes
    1.3.1—Insulation
    1.3.2—Drywall
    1.3.3—Interior trim
    1.3.4—Cabinetry
    1.3.5—Painting
    1.3.6—Flooring
    1.3.7—Finish HVAC, plumbing, electrical, and specialty work
    1.3.8—Trim finishes (door hardware, bathroom accessories, etc.)
    1.3.9—Phase inspection
1.4—Exterior finishes
    1.4.1—Final grade
    1.4.2—Concrete work (driveway, patios, etc.)
    1.4.3—Decking
    1.4.4—Guttering

> 1.4.5—Landscaping
> 1.4.6—Phase inspection
> 1.5—Final inspections
> > 1.5.1—Prefinal project manager inspection
> > 1.5.2—Final inspection
> > 1.5.3—Client punch list and repairs

Another level of detail is missing from this outline, but it will be included as each section is covered. Although the information is presented in a linear format, it might not necessarily be done in this order. However, it is necessary to present the information in this manner. For instance, guttering is covered much later in the outline than cabinetry is, but it is very possible that the guttering may be done weeks before the cabinetry is installed. The project schedule, not necessarily the flow of the WBS, will determine when exactly a portion of the WBS will be handled.

Section 1.1—Permitting and Site Prep

The first phase of the construction project entails completing all the paperwork, authorizing the construction work to begin, and preparing the site for groundbreaking. During this part of the project, the project manager is primarily concerned with making certain that the project moves through all the red tape as quickly as possible. Typically, this is a very straightforward process, which the project manager should be able to move through rather quickly. The project manager should still closely monitor the progress to make certain that everything progresses as planned. This section of the WBS includes the following parts (see Figure 4.1):

> 1.1.1—Prepare a site plan
> 1.1.2—Acquire permits
> 1.1.3—Site preparation

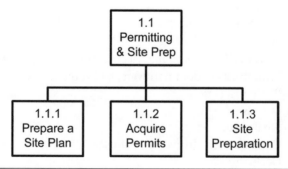

Figure 4.1 WBS Section 1.1

1.1.1—Prepare a Site Plan Preparing a site plan is one of the first steps of the execution phase of the project. It is possible that a generic site plan was prepared during the planning phase, but the project manager needs to make certain that the site plan adequately reflects the project before he begins the permitting process. Many times, building inspectors, town zoning agents, and erosion control inspectors will require a site plan before they will issue a permit. In some cases, a surveyor or engineer must be engaged to prepare the site plan, but this is not always the case. Sometimes the project manager feels he is able to do it himself. The typical steps to preparing a site plan are:

1.1.1.1—Review survey of lot
1.1.1.2—Walk lot
1.1.1.3—Draw detailed site plan
1.1.1.4—Flag desired house site
1.1.1.5—Flag septic field and other areas, if applicable

These steps can also be seen graphically in Figure 4.2.

1.1.1.1—Review Survey of Lot
First, the project manager needs to become familiar with the official survey of the property. A survey might be on file at the local register of deeds, or a recent survey might have been completed by the building company or the client. It should not be assumed that a current survey is available, as many areas do not require a survey for a new construction project. However, it is highly recommended that the project manager locate an accurate survey in order to pinpoint the corners of the lot.

1.1.1.2—Walk Lot
After becoming familiar with the survey, the project manager should take the survey to the job site to locate the corners of the property, as well as any other areas of note, such as streams, wooded areas, and any sloped areas. By walking the lot with

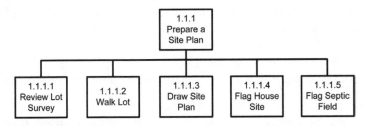

Figure 4.2 WBS Section 1.1.1

the survey, the project manager can make notes, which will be helpful in preparing the site plan. Time should be taken to note the natural lay of the land so that a few questions might be answered:

- Where will the water naturally drain?
- Are there any considerations due to the slope of the lot?
- What is the best possible location for the home?
- Where does the client want the home?
- If a septic system is necessary, where would the best areas be?
- Will the property need to be cleared of any timber or brush?

These questions have most likely been researched during the planning phase, but the project manager must make certain that he is knowledgeable about the building area and any natural or unnatural features that will need to be addressed during construction.

1.1.1.3—Draw Detailed Site Plan

After the project manager has reviewed the lot and compiled the notes, then she is ready to prepare a detailed site plan. Remember that the site plan represents the desired plan that the project manager is hoping will be approved. The various offices that have a hand in approving the construction project might make changes to the site plan, which will require it to be updated. The site plan should include at minimum the following items:

- Locate corners of property and distance between
- Locate and identify any roads
- Locate home site and dimensions
- Locate desired placement of septic field, if applicable
- Locate and measure driveway and any sidewalks
- Locate any geographical features: elevation variations, streams, etc.
- Determine distance between home site and edge of lot
- Establish building setbacks per zoning ordinances
- Locate any structures that shall be built as part of the contract (detached garage, guest home, sun room, gazebo, outdoor pool, etc.)
- Place directional arrow pointing North on map
- Include a scale of the plan
- Locate any easements on the property (water, sewer, power, cable, gas, etc.)
- Determine erosion control measures

The list is merely a guide. If the site plan is required for a permit, there might be other specific guidelines provided. Once the plan has been approved, the project manager should revisit the lot to verify that the site plan is feasible. The time should be taken to flag the home site and any other areas that will include improvements, such as a detached garage, the septic field, or the driveway. This way, when the inspectors arrive on the site, they will be able to easily determine the physical location of the various elements of the plan. If, during the preparation of this site plan, the project manager needs to make any modifications to the work completed during the planning phase of the project, he will need to update the relevant portions of the project's plan and update any stakeholders who need to be aware of the changes.

1.1.2—Acquire Permits Once the site plan has been prepared, the project manager may begin acquiring the various permits necessary for actual construction work to begin. Each county or municipality has its own forms and guidelines for acquiring the necessary permits, but the permits will, in general, include the following types, which are outlined in the WBS (see Figure 4.3):

1.1.2.1—Acquire zoning permit
1.1.2.2—Acquire septic system permit
1.1.2.3—Acquire erosion control permit
1.1.2.4—Acquire building permit

During the planning phase, it is likely that the project manager acquired blank copies of each of these permits so that she could familiarize herself with the requirements for each application. If this is not the case, then she might need to gather some documentation before she can apply. In some cases, this process might even require interaction with three or four different agencies. I work in an area in which a small town provides the zoning permit, a nearby larger town provides the septic permit, the county provides the building permit, and the state provides the erosion control permit. Therefore, the project manager needs to familiarize herself with the correct process.

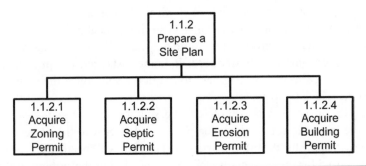

Figure 4.3 WBS Section 1.1.2

1.1.2.1—Acquire Zoning Permit

The zoning permit is typically the easiest permit to acquire. It will be given by the zoning department of the county or municipality that has jurisdiction over the areas in which the home site is located. The purpose of the zoning permit is to verify that the proposed building meets the various zoning ordinances that might apply to the property. The specificity of zoning ordinances and the input offered by the zoning department will vary by area. Many times, the project manager will simply need to complete an application and pay the fee, and the zoning permit will be issued immediately. In other areas, however, the zoning department might desire more input on the project. Zoning staff might wish to see the design of the home or a detailed site plan to ensure that the project not only meets the zoning ordinance, but meets their desires for the future growth of the community. For instance, if a builder were to attempt to build a very modern style home in a traditional downtown historic community, then she might meet with opposition on the basis that the proposed home does not conform to the area, and thus, the zoning permit might be denied. The project manager must change the plan, locate a new lot, or appeal to the town council, or the like.

1.1.2.2—Acquire Septic Permit

After the zoning permit has been acquired, the project manager will be ready to apply for a septic permit, if public sewers are not available. Applying for a septic permit typically requires the submission of a site plan showing where the project manager wants to locate both the home and septic field. After he makes the application and pays fees, he will schedule an appointment for a soil scientist to perform an inspection of the lot. Some environmental health departments require the project manager to prep the site before the inspection, which may include clearing wooded areas or brush and digging holes so that the soil is easily visible to the inspector. Some departments, however, do not require any site prep, and the inspector will come prepared to dig for soil samples.

The purpose of the inspection is to confirm that the property has soil suitable for a septic field. The quality of the soil and the area available for a septic field will determine how many bedrooms the home may have. If the soil quality is low, the number of bedrooms may be reduced. In some instances, the project manager might be required to hire an engineer to develop a solution if the county or city inspector is unable to do so.

After the inspection is completed, the soil scientist or engineer who performed the inspection must draw the plan and deliver it to the project manager before he can apply for the building permit. The plan will specify the type of septic system required for the project, as well as the location of the system on the lot. If the plan is different from the site plan prepared by the project manager, the project manager must update the stakeholders who need to be aware of the change, and he must update the project plan. Modifications that come back as a surprise might alter the construction cost, as well as the time required to complete the project. Therefore, it is important to begin addressing any changes as soon as possible.

1.1.2.3—Acquire Erosion Control Permit

The erosion control permit describes how the project manager will ensure that any erosion that occurs does not negatively affect the project. The offices that approve this are typically concerned with how the project manager will keep silt and run-off contained on the project site. This is typically accomplished through a couple of simple measures: silt fences and construction entrances. A silt fence should be installed on the low areas of the job site where the silt would otherwise run off onto either a neighboring lot or into a stream. Sand, rock dust, or gravel should be used for the construction entrance to the job site, which will keep the various vehicles from taking silt onto the highway.

Erosion control is becoming an area of major concern even in rural areas. In North Carolina, for instance, the state requires that job site reports be completed each week and after every major rainfall by the project manager to ensure that the erosion control measures implemented are working. The offices that manage this oversight will even perform surprise inspections at various times and provide a report of their findings. Noncompliance can lead to fines if not corrected. In most cases erosion control measures are relatively easy to maintain.

1.1.2.4—Acquire Building Permit

Acquiring the building permit will typically require:

- Building permit application
- Multiple copies of construction plans
- Copy of zoning permit
- Copy of septic permit
- Site plan
- Any engineering or survey reports

Once these items have been prepared, then the project manager should review the documents with a member of the building inspection office's staff to ensure that there is not any mistake or missing information that would create a problem. The time frame from submission of the plan to receipt of the permit will vary by area and depend on a variety of issues. The process typically takes only a few days, long enough for an inspector to review the plans to make certain all is in order. Just because a set of plans is approved does not necessarily mean that the plans will not have to be modified. If the building inspector missed confirming some aspect of the application, the builder might have to modify construction features even after the work has been completed. Therefore, if the builder is new to a particular area, she should make certain that all plans accurately reflect the appropriate portions of the building code. After a few days, the building permit will be ready to be picked up and posted on the job site. At this point, construction is ready to begin.

1.1.3—Site Preparation The first phase of actual construction is the preparation of the site. This process can take a few days, depending on the amount of preparation necessary. For instance, if the lot is heavily wooded, the site prep will take longer than if the home is to sit on a level, cleared lot. The typical steps of site preparation are (see Figure 4.4):

1.1.3.1—Install sedimentation control measures
1.1.3.2—Clear vegetation and debris
1.1.3.3—Do rough grading for house site
1.1.3.4—Install temporary power service
1.1.3.5—Inspect site work

1.1.3.1—Install Sedimentation Control Measures
The first step is to install the measures outlined in the erosion control plan. This will typically include installing silt fences and creating a construction entrance to the job site so that the amount of silt carried from the site is held to a minimum. The project manager wants to make certain that adequate measures are in place to ensure that, barring any unforeseen problems, the erosion control measures will remain effective for the duration of the project. Some project managers do the bare minimum, which causes the measures to fail. The work must then be redone. Once the erosion control measures have been implemented, the project manager needs to inspect the work to make certain that it was done according to the plan.

1.1.3.2—Clear Unwanted Vegetation and Debris
The grading subcontractor now arrives on the site and begins work. This grading is separated from the rough grading for the house site because sometimes two separate crews will be used for the work, as each job has its own concerns. This first time through is truly a rough grading job. The key here is that the subcontrac-

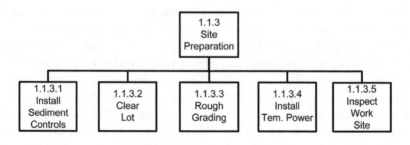

Figure 4.4 WBS Section 1.1.3

tor has the right equipment to do the work efficiently. Depending on the amount of land to be cleared, grading can take a few days and might affect the details of the contract. Most often, this first crew is concerned with removing most of the brush and trees, but they will not come back to remove the various tangles and roots that might still be protruding from the ground. That is typically done during the finish grade. After the site has been cleared, the project manager should inspect the work to confirm that it has been done in compliance with the contract.

1.1.3.3—Rough Grading for House Site

Many times this step and the previous one are done at the same time by the same subcontractor, but not always. The primary reason, however, that the two steps are separated is because of the different concerns associated with each task. In the previous task, the primary concern is removing the unwanted vegetation and trees in an efficient manner. In this task, the focus is preparing the land where the home will sit. This might require digging out a basement, resloping an area to correct any drainage areas, or a host of other concerns. This requires a skilled operator who understands how the topography of the land could adversely affect the home site, if graded incorrectly. The project manager should meet the grading subcontractor on the job site and discuss the project plan and the subcontractor's impression of the work. If the project manager is uncertain about a certain aspect of the work, he should ask for clarification. The two primary concerns are establishing a solid site for the home and making certain that rainwater will flow away before reaching the house site.

As the work progresses, the project manager should stop by and see the progression to make certain that the measures appear to be producing the desired effect. After the work has been completed, the project manager should perform a final inspection and approve payment for the work. As part of the contract with the subcontractor there should be a clause whereby the subcontractor agrees to return and repair any areas if the work was ineffective.

1.1.3.4—Install Temporary Power Service

At some point after the site has been cleared, the project manager will need to order temporary power service for the site from the local power company, so that the various workers will have access to power to operate their various tools. The electrical subcontractor commonly does this as part of the contract. Typically, power is installed near the front of the lot on a corner where it is not likely to be disturbed. The project manager should intermittently inspect the service to make certain that adjoining property owners are not availing themselves to the service. My company has paid the power bill for many motor homes over the years. When the owners of the motor home are questioned, they seem to have thought that it just a free service to the community by the builder. They did not know that anyone had to actually pay for the service.

Section 1.2—Foundation, Framing, and Rough-ins

Now that the appropriate permits have been acquired and the site has been prepared, the project manager is ready to authorize the actual construction work of the home to begin. This section of the WBS contains many of the most important parts of the construction project, as the work done now will affect almost every phase that comes hereafter. If mistakes are made here, they can create major problems throughout the remainder of the project. A footing that is uneven or out of square can cause problems for the foundation. A foundation that is off can cause the framing to be off, which can cause the drywall to be out of square, which can cause doors to be hung wrong, and the list goes on and on. Ensuring that work is done right now will make everyone's job easier later on in the project.

An important task that the project manager performs once actual construction begins is verifying that everyone knows what the plan calls for. Before any work is done by any employees or subcontractors, the project manager should meet with them briefly to review the details of their work to make certain that they are fully aware of the current status of the project and the details of their job. The project manager should make certain that the subcontractors do not have old specifications or the like. If there are any particularly challenging parts to their work, it might be a good idea to question them once again to make certain they understand the expectations clearly.

If the project manager will pay close attention to the details early on, the project team will make certain that they rise to the expectations that are set. If the manager begins allowing poor workmanship or delays, she risks the workers taking on this type of culture. She should spend more time following up early on to make certain that the workers understand that they will be held accountable for their performance on the project.

It is during this phase that the project manager begins working with inspectors on the job site as work is progressing. It is important to meet the inspector on the job site and show a willingness to work with them. The vast majority of inspectors are professionals who perform their jobs with excellence. There are, however, exceptions. If the project manager is working with an exception, he needs to know it as early as possible so that he can decide how he will deal with it. Most often, the project manager takes special steps to appease the inspector. If, however, the inspector cannot be appeased through careful management, the project manager might need to go to the inspector's supervisors. By doing this, the project manager is potentially putting himself in a difficult position, so this should only be done in extreme cases and with caution.

It is good for the project manager to show up at different times of the day to see the work in progress. For some workers, this will be an inconvenience, because the project manager may impede their work. These are the workers that do not need to be checked, but there are others who always seem to be starting back to work just when the project manager shows up on the job site. These workers need encouragement and unannounced inspections to encourage them to stay on task.

During this phase, the most visible work is being done. The clients will come to the job site one week and see nothing but a few block walls, and the next week, they will see a house. The project manager should tell them that the changes early on will be dramatic and seem to be progressing quickly, but that once the home gets under roof, progress will appear to them to slow down, as the changes are less dramatic. Part of the project manager's job is to manage expectations. The clients might think that if the work continues at this pace, they will be able to move in a couple of months early, so they need to be educated and updated on how the project will actually progress. It is not a bad idea to provide them with a construction flow chart. However, make certain that the dates are not included. If a construction schedule is provided to the clients, they will want to know about every deviation from it, so providing a detailed schedule is discouraged. If they insist on dates, milestones for each major phase of the construction workflow may satisfy them.

During this phase of the project, the project manager should keep a close watch on the project's adherence to the schedule, updating it as needed and then updating the project team and interested stakeholders regarding those changes. The project manager should stay in close contact with those subcontractors who will be expected to work on the project in the upcoming two or three weeks to make certain they arrive on time to begin their work on the project. If they cannot arrive on time, the project manager needs to know this as soon as possible to make certain that the plan is adjusted accordingly.

The major sections of this phase are:

1.2.1—Foundation
1.2.2—Framing
1.2.3—Rough-ins (electrical, plumbing, HVAC, etc.)
1.2.4—Exterior siding and trim

The sections can be seen graphically in Figure 4.5.

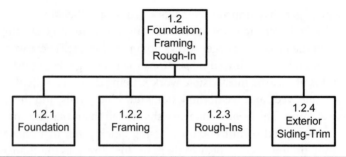

Figure 4.5 WBS Section 1.2

1.2.1—Foundation After the rough grading has been completed for the house site, the site is ready either to dig the footers in preparation for the foundation or to grade the site for the slab. The project example used in this book assumes that a footing will be dug for a foundation. During this initial phase of the project execution, the project manager will be communicating with various subcontractors and inspectors in something of a whirlwind of activity. Keeping a close watch on the project schedule and staying on top of communicating with the various stakeholders is critical to moving through this section of the project without losing any time. The typical process includes the following steps (see Figure 4.6):

1.2.1.1—Stake and dig footing
1.2.1.2—Footing inspection
1.2.1.3—Pour footing
1.2.1.4—Build foundation

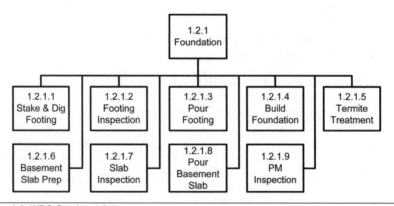

Figure 4.6 WBS Section 1.2.1

1.2.1.5—Termite treatment, if needed

1.2.1.6—Basement slab preparation

1.2.1.7—Slab inspection

1.2.1.8—Pour basement slab

1.2.1.9—Project manager inspection

1.2.1.1—Stake and Dig Footings

Knowing where to position the home is critical to this first step. The home site has been flagged as part of a previous phase, but those flags have most likely been moved for the grading. The project manager will need to use the site plan to reflag the house site. After the home site has been flagged, the subcontractor or workers are ready to stake out and dig the home's footings. The design of the home will specify the appropriate type of footing, which the project manager will need to verify with the subcontractor before work begins to make certain that everyone understands the requirements of the project.

For an average-sized home, most footings can be dug in one day. For smaller homes, the digging and pouring of the footing can take place possibly on the same day with good planning. But in this example, a lag will be assumed between the two events. However, this not always the case. After the footing has been staked and dug, the project manager will want to verify the dimensions, depth, and accuracy of the footing by comparing it to the specifications of the project plan. Once the footing has been dug, an inspection is typically necessary.

1.2.1.2—Footing Inspection

The local building inspection office will perform the inspection. Depending on the notice required to schedule an inspection, the project manager may be able to call one day prior to the date that the footing is scheduled to be dug, or he may need to schedule a few days in advance. Earlier in the project plan, the need to become familiar with the local inspectors and their standard practices was discussed. In doing so, the project manager can know when to schedule the various inspections in order to have the work completed according to the project's schedule.

It is a good idea for the project manager to meet the inspector at the job site for this initial inspection to make sure that a good relationship is developed from the beginning. The inspector will be checking the depth of the footing, the compaction of the soil, and overall design of the footing to make certain that it complies with appropriate building codes. If the inspection can be scheduled for a time shortly after it has been dug and while the subcontractor is still on the job site, the building inspector may allow the subcontractor to quickly fix any issues that arise. This will allow the footing to pass on the first inspection. If this is possible, it could save a couple of days.

1.2.1.3—Pour Footing

After the footing passes inspection, then it is time to pour the footing with concrete. This process will typically only take half a day, depending on the footprint of the home. The project manager should ensure that those who pour the footing make certain that the concrete is level, so that there will be no major issues for the masonry subcontractor. As a teenager, I worked for a masonry subcontractor over a summer or two, and there were few things that frustrated the crew more than an unevenly poured footing.

After the footing is poured, the project manager should perform a final inspection to make certain that the work is in compliance with the specifications of the project plan. If the work is approved, the project manager should authorize payment and see that the job site is prepared for the masonry subcontractor.

1.2.1.4—Build Foundation

This step includes both ordering materials and construction of the foundation for the home. During the planning phase, the masonry subcontractor most likely provided an estimate of the amount of materials needed for the job. The materials will be either cinder block or a precast foundation system. The project plan will specify which type of foundation the project requires. If the project calls for the precast foundation system, the system must be preordered so that it can be constructed. The delivery of the system and the installation should be closely coordinated with the project schedule to avoid any delays. The installation of this type of system is typically relatively quick, as long as all goes according to plan.

If the project, however, calls for a more traditional block foundation, the project manager will need to have the materials delivered to the project site. The materials necessary for the project will be detailed as part of the project plan. The subcontractor typically provides the materials list. Once the materials are delivered, the project manager should verify that the correct materials were shipped and review the project plan with the masonry subcontractor just before work begins on the foundation to ensure that the plan is clear.

Laying a traditional block foundation may take anywhere from a few days to well over a week, depending on the size of the project. As work progresses, the project manager may choose to drop by to to make certain work is progressing according to schedule. If she sees the work begin to lag behind, she should confer with the masonry subcontractor as to whether the schedule needs to be revised due to weather or other factors.

The masonry subcontractor is also responsible for any foundation waterproofing that is necessary for the exterior foundation wall. Waterproofing is required if any portion of the foundation will be underground. It is imperative that this is properly done, as a poorly water-proofed basement can create numerous problems in the future. The project manager should also make certain that the foundation drain is installed correctly.

Once the foundation work is complete, the project manager should inspect it to verify its compliance with the project's specifications. Any extra materials should be catalogued. If they can be returned, then they should be. If not, the project manager should see if she can use them later in the project. Damaged or extra materials should never be given to anyone, as it creates an incentive to damage materials or to have extras at the end of the job so that the workers can have the benefit of free materials for their own personal projects at home.

1.2.1.5—Termite Treatment (if needed)

After the foundation has been built, the home should be sprayed for wood-destroying insects if the local building code requires. This is a requirement of the building code in many, but not all areas, and a licensed company must perform and certify the work with a warranty. Typically, the warranty will be for one year; the homeowner can usually extend the warranty by signing a service contract with the treatment company.

1.2.1.6—Basement Slab Preparation

If the home has a full or partial basement that calls for a concrete floor, then some prep work must be done before the slab can be poured. In many cases, this work is done by the concrete company that will be pouring the basement slab. In other cases, it may be necessary to hire separate subcontractors. One company will prepare the area for concrete, and a second company will pour and finish the concrete.

The purpose of the preparation is to create an appropriate and level base onto which to pour the concrete. A combination of sand, gravel, and rebar is typically used to reinforce the concrete. Sand is an excellent base, as it will compact very well, preventing future cracks in the concrete. After the area has been prepared, the project manager should make certain that everything has been properly completed in preparation for the inspection.

If there is to be any plumbing or lines run under the concrete floor of the home, now is the time to install them. Any drains in a garage, plumbing for a bathroom or garage sink, or a drain line for the septic tank should be installed by the appropriate subcontractor and inspected as part of the slab inspection. In this process, the spacing of these tasks is critical. If the plumbing is installed in the wrong location, the concrete will have to be cut and patched at a later date, which can slow the progress of construction. The project manager should take the time to ensure that the lines are installed in the correct location.

1.2.1.7—Slab Inspection

After the area has been prepared, another inspection is required. The building inspector will make certain that the ground is properly compacted and that any plumbing or drain lines have been properly installed. The project manager should meet the inspector on the job site with the subcontractor to make certain that if the work fails the inspection, the subcontractor understands why. If the failure is due to a small issue, the subcontractor may be able to correct the work immediately so that the inspector will approve the work.

1.2.1.8—Pour Basement Slab

After the inspection has been passed, the basement floor can be poured. This typically requires less than a day, depending on the size of the basement. The project manager determines whether the current weather conditions are conducive to the work being done. If the home is being built in a colder climate, the project manager should ascertain whether insulation pads will be necessary to protect the concrete until it has stabilized.

1.2.1.9—Project Manager Inspection

The project manager is advised to inspect the work as it progresses through each step. However, there are certain milestones at which a project manager inspection is especially important. This ensures that the project does not fail in any area that might create a larger concern later on if uncorrected. The project manager should review all the foundation work to date to ensure its compliance with the project plan. He should also take the time to review the project plan and the progress to date and update the project schedule, reflecting any delays or changes between the planned and the actual timeline.

1.2.2—Framing After the foundation has been completed, framing can begin. This phase may include framing the actual exterior walls of the home, or it may call for framing the interior walls of a block home. For the sake of this discussion, however, let's assume that the above-basement levels of the home will be wood framed. It is helpful to think of framing in a couple of sections or parts. The WBS used in this example has the following steps:

1.2.2.1—Order framing and roofing material package
1.2.2.2—Frame floor and wall systems
1.2.2.3—Install or build roof system
1.2.2.4—Install windows and exterior doors
1.2.2.5—Framing and window inspection
1.2.2.6—Project manager inspection

The project manager might modify the breakdown of tasks to meet particular needs. A sample of the above WBS can be seen in Figure 4.7.

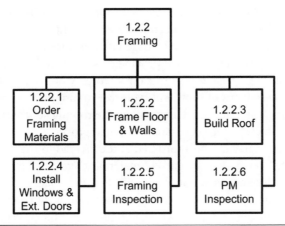

Figure 4.7 WBS Section 1.2.2

1.2.2.1—Order Framing and Roofing Material Package

The material order for framing the home is one of the largest material orders in the construction project. It will include:

- Framing materials
- Roof truss system
- Floor truss system
- Roofing materials

The project manager should complete the framing material order list. If he does not possess the expertise to create the framing materials list, he can ask the framing subcontractor to create the list. Another option is to have the building supply store create the list from the blueprints. In any event, the framing subcontractor should at least review the materials list before it is ordered. Unless some unique materials are specified, most will be kept in stock at the building supply vendor, and they can be delivered to the job site within a day or two of the order being placed.

The project manager must also order the truss systems for the roof and floor. This might require more lead time than needed for the framing materials, unless the vendor has the product needed in inventory. Typically, the project manager sends the plans to the truss vendor, and he will return a list of recommended materials based on the project specifications. Ideally this takes place during the planning phase of the project; if not, it must be done at this point in the project. The time required to build the truss systems will vary based on a variety of factors. Once the systems have been ordered and delivered, the project manager should make certain that the installation follows the guidelines provided by the truss vendor. A copy of the truss specifications and installation instructions must be forwarded to the local building inspector's office to be filed with the project files.

Roofing materials are the last to be ordered in this. If the framing subcontractor is also doing the roofing, she will provide a list of materials. The home might call for a shingle roof, a slate roof, a metal roof, or any one of a host of other roof types. Some of these types of roofs will require a roofing subcontractor who specializes in certain materials. It is best to allow the roofing subcontractor to create a materials list, which the project manager then approves.

After the materials have been ordered and delivered to the job site, the project manager must verify that the correct materials were delivered. He should not only check quantity and type, but also verify that the quality of the materials is acceptable. A certain percentage of the materials will likely need to be returned for replacement due to bows in the wood, and the like. If the project manager feels unqualified to perform this type of inspection, he should have another qualified member of the project team perform the inspection.

1.2.2.2—Frame Floor and Wall Systems

The subcontractor builds the flooring system either by constructing one or by installing a floor truss system. The choice is based on the needs of the project. Typically, if the client wants more clearance on the lower level for the purpose of a recreation room, or the flooring system is to be between floors in a two-story home, then the floor truss system is the preferred option.

Once the flooring system has been installed, the framing subcontractor constructs the wall system of the home. As this work is done, the project manager should verify that the layout conforms to the blueprints by checking room dimensions. There may be some minor variation of an inch or two, which is acceptable, but the actual framing should conform very closely to the blueprints.

1.2.2.3—Install or Build Roof System

Once the flooring system and walls have been framed, the next step is to set the roof trusses or to manually construct the roof system. Most home designers will use a roof truss system because of the many design advantages. Often, a roof truss system will remove the need for load-bearing interior walls, giving the client greater flexibility in space design. However, at times, it is more advantageous to stick-build a roofing system because of a particular design feature of the plan. The choice of method should be based on expert advice from the framing subcontractor, an engineer, or a project architect.

After the roofing system has been completed, the roofing subcontractor will come to install shingles, metal, or some other material that will be used on the roof of the home. Many times the workers will be the same crew that is framing the home. If it is a different subcontractor, however, the project manager might find it more helpful to divide this part of the WBS into two sections.

After the framing and the roof have been completed, the project manager should perform an inspection to make certain that the work to date is in accordance with the project plan. She should also confirm that the windows are located in the correct position before they are actually installed, which is the next step.

1.2.2.4—Install Windows and Exterior Doors

Once the project manager has confirmed the placement of the windows and exterior doors, the framing subcontractor can install them. Care should be taken to make certain that the windows are without defect at the time of installation. Although changing out windows is not an impossible task by any means, it is a hassle that should be avoided if possible. Therefore, the short time that an inspection will take is well worthwhile.

As part of the installation, the windows and doors will be wrapped. This refers to installing a barrier of felt or other such material around the windows and door frames according to the construction code. Sometimes the entire home will be wrapped, but in most places, the building code only requires wrapping for windows and doors.

1.2.2.5—Framing and Window Inspection

Once the framing is completed and the windows and doors have been installed, the home will most likely need to be inspected by the local building inspector. The inspector makes certain that the framing is completed in a manner that complies with the local applicable building code and that the windows are properly installed. As before, the project manager should meet the inspector for the inspection to make certain that any corrections or modifications are clearly understood. If rework is needed, the work will need to be scheduled, and reinspected.

1.2.2.6—Project Manager Inspection

Before the project manager progresses to the rough-in stage, she should make a final inspection of the property to make certain that no modifications are needed. If the home has already been sold, it will be a good idea to walk through the property with the client to make certain that all is as expected. At this point, modifications are much less expensive than they will be in coming days.

The project manager should also take the time to make certain that the project file is up to date. She should review the budget, the schedule, the risk management plan, and the other relevant sections of the project plan to ensure that updates have been made as the project has progressed. As well, the various team members, subcontractors, vendors, and other relevant stakeholders should be updated as to any changes in the project plan.

1.2.3—Rough-ins The clients have just witnessed the most dramatic visual progression that occurs during the construction project. As they see the home framed, it will seem to them as though the project is progressing along very quickly. They might think that the home will even be finished a month or so early because of the dramatic progress. Then comes the rough-in stage, which is without a doubt one of the most boring portions of the project for the clients. They will come in and see some plumbing pipes installed or a few trunk lines installed, or even some wiring run, and they will wonder what is taking so long. Is no one showing up? Did the project manager forget about the work?

This will be the perspective of many clients, but nothing can be further from the truth. The product of the rough-in stage can be compared to the circulatory or nervous system of body. Individually, each bit seems mundane and unimportant, but getting this bit right will ensure a home that meets both the needs and desires of the clients. If this portion is poorly done, it creates many headaches and future frustrations. Time must be taken to ensure that the work progresses according to plan.

The typical steps associated with the rough-in stage are (see Figure 4.8):

1.2.3.1—HVAC rough-in
1.2.3.2—Plumbing rough-in
1.2.3.3—Electrical rough-in
1.2.3.4—Specialty rough-in
1.2.3.5—Project manager inspection
1.2.3.6—Rough-in inspections by building inspector

Before each of these steps is discussed, a point on order should be made. The order listed above reflects the most typical order, but at times it may be necessary or desirable to deviate from this sequence. The project manager should look at the specific needs of the project to determine which order would be most desirable.

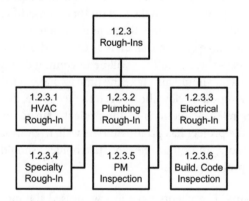

Figure 4.8 WBS Section 1.2.3

1.2.3.1—HVAC Rough-in

The HVAC (heating, ventilation, and air conditioning) system typically can be completed in a day or so depending on the size of the home. Typically, the HVAC subcontractor will run trunk/ventilation lines through those areas of the home that he will not be able to access once the drywall has been installed. He will also cut holes in subflooring to identify where the vents will be in the floor.

Once the subcontractor has finished the work, the project manager makes certain that the placement of the trunk lines and floor and ceiling vents comply with the project plan. Once the work has been completed, a partial payment is typically made, but the majority of the payment is withheld until after the subcontractor completes the work later in the project.

1.2.3.2—Plumbing Rough-in

The plumbing rough-in entails installation of any prefabricated showers or tubs, as well as running the water and drain lines. Typically, the plumbing subcontractor will be responsible for ensuring delivery of the prefabricated bathtubs according to the subcontractor contract. If this is not the case, the project manager must ensure that the right materials are delivered to the job site. The project manager should review the project plan with the plumbing subcontractor to ensure that the water lines and the drain lines are installed in the correct location in the kitchen and bathrooms and utility room. If the home requires a private septic system or hooks into a community or shared septic system, then the project manager must provide the appropriate paperwork to the plumber to make certain that the drain lines are run to the correct location to be properly connected.

After the plumber has run the various water lines and drains, the project manager should perform an inspection to make certain that all has been installed in accordance with the project plan. Typically, the subcontractors that perform rough-ins are paid a percentage when they rough-in the home and the balance when they return and complete the work, which is called the *finish*.

1.2.3.3—Electrical Rough-in

In most cases, the electrician prefers to be the last subcontractor to do the rough-in, although this is not required. The electrical system is one of the more critical systems, which, if planned well, has the potential to greatly enhance the owner's enjoyment of the home. A poorly planned electrical design can lead the homeowner to conclude that the electrician was incompetent, even though the actual quality of the wiring is first rate. Light switches, outlets, telephone lines, and cable outlets must be thoughtfully placed throughout the home to meet the needs of the client. Great care should be taken to ensure that the electrical system accommodates the lifestyle and needs of the client.

After the electrical wiring has been completed, the project manager should ensure that the work is in accordance with the project plan. An important point to note here is that the electrical subcontractor might wish to coordinate the work not only with the plumber and the HVAC subcontractor, but also with the subcontractor installing the exterior siding on the home. If the exterior siding of the home requires a material that will either be nailed or screwed, the electrician may want to wait until after the exterior siding has been installed before completing the rough-in. In some circumstances, a nail can split a wire in the wall, which is not discovered until much later in the construction project when the power is turned on to test the electrical system.

1.2.3.4—Specialty Rough-ins

The three rough-ins discussed so far are common to every construction project, but there are additional rough-ins that may be necessary. On the sample WBS a few examples are listed:

1.2.3.4.1—Central vacuum
1.2.3.4.2—Alarm system
1.2.3.4.3—Television system
1.2.3.4.4—Home networking system
1.2.3.4.5—Audio visual system

If any such system is specified in the plans, it must be roughed-in prior to the insulation being completed. Depending on which options are selected by the client, the project manager will need to oversee the work and make certain that it is done in accordance to the client's desires. In some cases, the local building inspector may be unfamiliar with the specifics of a certain system; the vendor must provide installation specifications for the building inspector to review. In some unusual cases, the vendor who has installed the system may need to be on the job site for the inspection to answer any questions that the inspector may have.

1.2.3.5—Project Manager Inspection

After the rough-ins have been completed, the project manager should perform an inspection prior to ordering the rough-in inspection from the building inspection office. Although the project manager might not have the knowledge necessary to determine whether the work will pass the rough-in inspection, she should be able to review the work in general to confirm its conformance to the project plan.

1.2.3.6—Rough-in Inspections by Building Inspector

If the project manager has taken the time to review the requirements for the rough-in inspection with the local building inspector and has communicated those requirements to the subcontractors, it is likely that the inspection will pass the first time. Once again, it is a wise idea to have a member of the project team at the job site to meet the inspector to ensure that any directives to modify the work are correctly understood. After the inspection is completed and any necessary rework is completed and reinspected, the interior portion of the work progresses to the insulation stage.

1.2.4—Exterior Siding and Trim Once the home has been framed, work can begin on the exterior of the home. There are a number of exterior materials available ranging from masonry related (brick, rock, etc.) to vinyl, composite, or even natural (wood) siding. However, when the project plan for the home is developed, a type of exterior finish is selected according to either the customer's or the project sponsor's choice. Therefore, the project manager will simply be implementing whatever selection has been made.

Before the material is ordered and installed, the project manager should review the project plan to be certain that the material selected is in conformance with any deed restrictions or the like. Many times, a set of neighborhood building guidelines will be recorded and referenced in the deed. These specifications will either specify certain materials to use or materials not to use. If this is overlooked, and the project manager has the incorrect materials installed, it will be considered his fault, and it can be a very costly mistake.

The title of this section of the WBS is called exterior siding and trim. Exterior siding refers to the main material that is used on the exterior of the home, and trim refers to any minor materials that might be used to complement the primary type of siding. Distinguishing between the main exterior siding and the exterior trim is useful as an example, but it might not be useful enough for one's specific project. This is because one may be using multiple materials (vinyl siding, brick, rock, etc.) that require unique planning for each material. If this is the case, the project manager needs to modify this section to fit his particular needs.

This example draws a distinction between the siding used on the foundation and the siding used on the main levels of the home, but this distinction will not always be helpful, which can be seen in Figure 4.9. For instance, when brick is used as the siding for the main level of the home, it is typically also used for the foundation as well, and both the foundation and the main level are done at one time. Another area where the separation used in the example will not be helpful is when multiple types of siding are used that require different subcontractors. For instance, part of the siding might be brick or stone, whereas another section is vinyl siding or a wood siding. The project manager will want to design this section of the plan carefully to make certain that the distinctions made are helpful and apt for the current project.

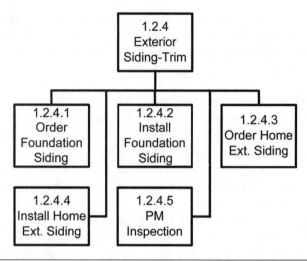

Figure 4.9 WBS section 1.2.4

The sections identified in the example WBS are:

1.2.4.1—Order foundation siding
1.2.4.2—Install foundation siding
1.2.4.3—Order home exterior siding
1.2.4.4—Install home exterior siding
1.2.4.5—Project manager inspection

1.2.4.1—Order Foundation Siding

The project manager's first step is to place an order for the siding that will be used on the foundation. The timing for this is dependent on the lead time necessary for the materials to arrive and the date that the work is scheduled to begin. Typically, one does not want material sitting on the job site for too long unless there is a way to secure it. The good news about this type of material is that it is typically heavy, so it is not easily stolen. If the job requires bricks or stones, the project team should not be surprised if a few are lost to the neighbor's outdoor landscaping project.

When placing this order, the project manager should carefully consider where the materials should be placed on the job site. Many times, the delivery team will simply place the material in the easiest place to get to, which is not always the most convenient place for the crew installing it. The project manager should check with the subcontractor who will be installing the material to determine where they want it placed on the job site, and then communicate this information to the delivery crew.

Once the order arrives on the job site, the project manager should have it checked for accuracy, with reference to quantity, type, and color. A problem can arise here with natural materials that the client has selected. Often, clients will select a natural stone for the foundation based on a few examples they have seen, but they might not be aware of the full extent of variety that can exist within the material selected. The project manager is wise to make certain there are not any specific rock types that the client does not want on the home, so that the subcontractor will not use them. This sounds like a lot of hassle—because it is—but it is more work to take the stones off the foundation and replace them.

1.2.4.2—Install Foundation Siding

After the siding has been delivered and all preceding work has been completed, the project manager is ready to order that work begin on installing the siding on the foundation. Most often this work is done by a subcontractor, so the project manager's role is one of supervision from a distance. Checking back at the job site while work progresses is not a problem, but the project manager should not try to run the subcontractor's crew for him. Checking in every so often will allow the project manager to confirm that work is progressing as it should, in reference to both time and quality.

1.2.4.3—Order Home Exterior Siding

Ordering the exterior siding for the main level of the home is presented here as a separate step, when in fact it might not be. The siding used on the exterior walls of the foundation might be the same as used on the exterior of the main levels of the home. For the sake of the example, let's assume that they are different.

Here, as with the foundation siding, the project manager's concerns are to ensure that the siding is in compliance with any restrictions or guidelines that might be governing the property and that the materials are ordered in a timely manner. Once the materials arrive on the job site, the project manager should either inspect them personally or have someone inspect the order for accuracy. If this is not done, and the wrong materials are installed, it can be a costly mistake.

1.2.4.4—Install Home Exterior Siding

As with the installation of the foundation siding, the project manager's role in this phase is primarily that of supervising from afar. The difference with the exterior siding of the main level, however, is that, more often than not, multiple types of siding are used. If this is the case, or if there are any special instructions the client wants followed, the project manager should take the time to confirm that the subcontractor has a clear understanding of what the client's expectations are. The problem that some project managers can run into in this situation is to assume that everyone has a shared understanding of the specifications. This might not be the case. The project manager must confirm that both the client and the subcontractor are on the same page as the project manager.

Once this is done, work should commence. Seeing the exterior siding installed is an enjoyable time for most parties, as it really brings together the look of the home. The client can finally see how the home is materializing. Now the inside is still rather skeletal, but the outside should be starting to look rather complete. Because of this, the project manager must pay close attention to the client's expectations. As the client sees the exterior of the home come together seemingly quickly, she might assume that the interior might come together just as fast, which is not often the case. The project manager should make certain that unspoken, unrealistic expectations are not entering her mind at this point. This can be done in an indirect manner by reviewing the progress so far and making comments about the upcoming work and the time needed to complete the remainder of the project. Even though the project manager quoted a specific amount of time, clients have a way of reducing it in their minds as the project progresses.

1.2.4.5—Project Manager Inspection

Although the project manager has been inspecting from afar to see how the work has been progressing, he should still perform a final inspection of the work done to the exterior of the home. At this point, he is simply verifying that the work performed is in conformance to the work planned. If there are any special notes or areas of concern that the client had, the project manager pays close attention to those items.

Section 1.3—Interior Finishes

This is a rather broad category within the WBS that covers a range of work: insulation, drywall, interior trim, cabinetry, painting, flooring, finishes on the HVAC, electrical and plumbing, and trim finishes. A sample WBS is shown in Figure 4.10. Balancing all the workers and orders that are necessary to successfully navigate this phase of the project can be quite challenging. Taking the time to oversee all this work can mean numerous trips to the job site to meet with the subcontractors, inspect material deliveries, and oversee the progression of work. Timing and communication becomes increasingly critical, as more and more workers need access to the job site to complete their portion of the project. In the following discussion of each major step in this phase, the specifics of construction will not be discussed as much as the methods of managing each section.

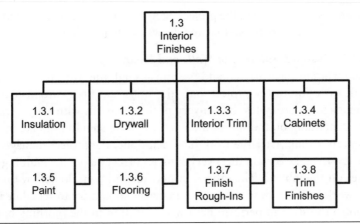

Figure 4.10 WBS Section 1.3

1.3.1—Insulation The insulation process consists of at least three steps (see Figure 4.11):

1.3.1.1—Order insulation materials
1.3.1.2—Install insulation in walls and floors
1.3.1.3—Inspect insulation

1.3.1.1—Order Insulation Materials

The WBS dictionary should state what the lead time on delivery of the insulation materials is, as well as specify what materials need to be ordered. The project manager ensures that the materials arrive on time, which means she coordinates with the subcontractor or company employees installing the insulation. The material must be stored in a secure and dry location until it is installed. Once the materials are delivered, the project manager inspects the order and makes certain that it

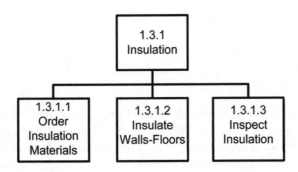

Figure 4.11 WBS Section 1.3.1

matches what was specified in the project plan. If there is a deviation, then she will correct it as soon as possible so that any delays can be avoided.

1.3.1.2—Install Insulation in Walls and Floors

After the materials have arrived and been inspected, the work is ready to begin. During this portion of the project, only the walls and floors are insulated. The project manager coordinates the delivery of the materials with the arrival of the workers. Depending on the size of the job, insulation should not take more than a few days to install. After the work has been done, the project manager does a quick review to make certain that everything is correct before the building inspector arrives on the job site. The official inspection will typically follow on the day after the work is finished, so the project manager should try to schedule it in the afternoon so the she can perform her own inspection in the morning before the inspector arrives.

1.3.1.3—Inspect Insulation

The insulation inspection is a routine process in which the inspector makes certain that the right type of insulation has been used and that it has been installed properly. Typically there are no surprises. Just in case, however, the project manager should do a brief inspection before the inspector arrives to make certain that no insulation has been removed or damaged between the time that the work was finished and the inspector arrives.

Many times, the same subcontractor will be insulating the ceiling, which comes later in the project. If this is the case, a partial payment on the contract is typically made at this point.

1.3.2—Drywall The next work to be done on the interior of the home is the drywall (see Figure 4.12). Drywall is called by a number of names: gypsum board, plasterboard, wallboard, and a few others names depending on the region in which the construction is being done. Drywall is used for the finish construction of walls and ceilings. The steps to this section are:

1.3.2.1—Order drywall materials
1.3.2.2—Hang drywall
1.3.2.3—Finish drywall
1.3.2.4—Contractor/project manager inspection
1.3.2.1—Order drywall materials

Typically, a big lead time for the delivery of drywall materials is not necessary. The WBS dictionary should specify the number of boards and the types of boards necessary for the walls and ceilings. In my experience, the contractor provides the drywall, whereas the subcontractor provides the joint compound and tape necessary to finish it. Once the delivery arrives, the project manager should make cer-

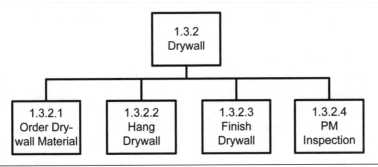

Figure 4.12 WBS Section 1.3.2

tain it is inspected for accuracy. It is even a good idea to show up or have someone there to make certain that the delivery team does not simply stack it in the garage or the basement; instead, they should spread it out throughout the home to make it easier on the installation team.

A special note to remember is that, in most cases, the drywall delivery is the last heavy shipment that arrives on the job site. My company typically uses this as a signal to begin working on some of the exterior projects, such as the driveway and finish grading. Other deliveries will come, but in most cases, they would most likely do little to no damage to any exterior work done.

1.3.2.2—Hang Drywall
Once the materials are delivered the drywall is ready to be hung. But one important step should be taken first: a brief inspection. Recall that after the framing crew finishes their work, the project manager should inspect it to make certain that the floors are level and the walls are square. This way any deviation could be corrected early on. Just before beginning the drywall installation, the project manager should do another quick inspection. Once the drywall is hung and finished, it is much more work to correct any problems that might be discovered when the trim carpenter arrives.

If it is possible, it is best to have the same subcontractor do the drywall hanging and the finishing. The reason is a practical one. If the finish work looks sloppy in some area, the subcontractor that is finishing it will say that the drywall was hung incorrectly. When the workers who hung it are called back to the job site, they will claim that the guys who are finishing it just do not know how to do their job. Of course, when the same crew does the hanging and the finishing and a problem arises, they will most likely blame the framing crew.

Hanging the drywall will go relatively quickly, depending on the size of the home and the size and ability of the hanging crew. A good crew will spend more time laying out their plan for hanging, which will result in less waste. The project manager should check the work early on to ensure they are using the materials wisely.

1.3.2.3—Finish Drywall

Finishing the drywall takes more time than hanging it, because time is needed for the joint compound to dry and be sanded. If the weather is especially humid or damp, it can delay the drying time, which in turn can lead to delays in the project. The project manager must keep a close watch on how the work is progressing to make certain that the schedule is updated and that the subcontractors who follow are cued to any schedule changes. During this couple of weeks, it is a good time for the project manager to move his attention to exterior aspects of the project, which will be discussed later.

1.3.2.4—Project Manager Inspection

The project manager is encouraged to come by and visit the job site while the dry-wall work is being completed. After the work is completed, he should complete a more formal inspection before the trim work is done. He should pay special attention to two items: (1) corners and edges and (2) smoothness. Using a level and a square, he should check doorways, corners, and any other edge that he can find to ensure that all is as it should be. In addition to making certain that the joints are square, the project manager should also verify that they are straight. This is often a problem with cathedral ceilings.

The second area that should be inspected is the smoothness of joints. If screws were used, they are less likely to be pushing out; but if nails were used, this problem is more likely. If the drywall is not smooth, this will create problems when the home is painted—especially if a satin or semi-gloss paint that has a sheen is used. If the project manager discovers problems after the paint is applied, then a lot of sanding and repainting the wall will be required. If deep colors are used on expansive walls, rework can be especially problematic.

1.3.3—Interior Trim

Trimming out the home is a very exciting time on the job site. After the drywall is installed, the home takes on an entirely different look for sure, but after the trim is done with doors hung, the baseboards installed, the crown molding put up, and the windows trimmed out, everything really seems to come together. The trim work can be done all at once, but it is usually done in a couple of stages, depending on the needs of the project. The reader should remember that we are moving through the project based on the WBS. This is not necessarily the exact order that the project will be scheduled. Many times, if the cabinets are ready after the drywall is completed, they will be set in place before the trim work is done. If they are not ready, the trim elsewhere in the home will be done. The trim in areas that call for cabinets, including custom entertainment centers and bookshelves, will have to wait. These areas might include the kitchen, bathrooms, mudroom, laundry room, and so on. Another consideration before the baseboards are installed is the type of flooring to be put down. For instance, if hardwoods and tile are going to be installed, part of the trim work may be done, but the baseboards will be delayed until the flooring is installed. Then the baseboards can be set directly on the hardwoods so that no shoe molding will have to be used; this gives a nice, clean look. The interior trim section, at minimum, consists of (Figure 4.13):

1.3.3.1—Order trim package
1.3.3.2—Install trim package
1.3.3.3—Inspect work

1.3.3.1—Order Trim Package

The trim package is typically ordered as early as necessary to get all the materials to the job site on time. There are a few considerations to take into account. First, the project manager should see if there are any special order items that are not stocked at the building supply company. Special order items may require more preplanning. For instance, many building supply companies only hold a small supply of interior doors that come with brushed-nickel door hinges. Instead they will typically stock doors with the standard polished-brass hinges. To avoid changing out the hinges on every door for the customer, the project manager should build in some extra lead time to order the correct doors. Second, the builder needs to know how the work will be done. Will it be done all at once, or will it be done in phases? If it will be done all at once, then all the materials can be delivered initially. If, however, it is going to be done in phases, only a partial order may need to be delivered, unless the project manager has a safe and dry place to store the materials where they will not hinder other workers.

Typically, the trim carpenter will have provided the project manager with a list of necessary materials during the planning phase. However, the project manager

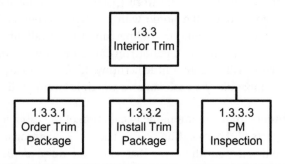

Figure 4.13 WBS Section 1.3.3

must review this list for accuracy before the order is made. If anything has been modified during the previous construction work, which has caused a deviation from the original project plan, this will need to be accounted for when ordering the trim. Again, when the order is delivered, the project manager should have it reviewed not only for accuracy, but also in terms of the quality of the materials. If the chair rail or the crown molding is damaged or warped, it will need to be replaced. Doors also need to be checked for scratches and broken jams. Taking the time to do this when the materials are delivered will ensure that the work is ready to begin when the trim carpenter arrives.

1.3.3.2—Install Trim Package

Once materials arrive, the trim work is ready to be done by the trim carpenter. In the couple of weeks leading up to this, he should be contacted to make certain that the scheduling will correspond with the needs of the project. If he arrives a couple of days before the cabinetry has been installed, he may work toward that portion of the home. If he arrives after the cabinets have been installed, he can install trim around the cabinets.

Earlier it was stated that the project manager must inspect the drywall after it has been finished to make certain there are no problem areas. Before the trim carpenter begins, the project manager should also check over the home to make certain there are no areas that will create any problems. Once the crown molding has been installed, it becomes very obvious if the ceiling joint is uneven; this can be avoided through a careful inspection of everyone's work.

The type of flooring to be installed in the different rooms throughout the home determines if the trim carpenter will finish work before or after the floors are installed. If tile or any type of hardwood flooring is being installed, it is best to wait on the baseboard trim so that it can be set directly on the flooring. If carpet is being installed, the trim carpenter can install the trim without any concern.

1.3.3.3—Inspect Work

After the trim carpenter finishes the work (partially or completely), the project manager should go through the home with the trim carpenter inspecting the tightness of the joints, the straightness of lines, and the overall quality of the work before authorizing any payment. A good trim carpenter will not mind this, as he knows that his work will endure a little scrutiny. If the trim carpenter becomes defensive, there is most likely a problem area which is awaiting discovery.

1.3.4—Cabinets In my opinion, the installation of the cabinetry is one of the most enjoyable parts of the execution phase. There is something about seeing the cabinets in place throughout the home that sends the *we're almost finished* message. The right cabinets and countertops can really bring the home together; they should complement and interact with the flooring and the paint choices. They can also be integrated into the home through the ceiling designs and wall textures. The choice of materials is part of the planning phase; now the project manager gets to see if it looks as good in the house as it did on paper. The steps associated are (see Figure 4.14):

1.3.4.1—Order cabinets
1.3.4.2—Build cabinets
1.3.4.3—Install cabinets
1.3.4.4—Inspect cabinets

1.3.4.1—Order Cabinets

The first step is to order the cabinets. A great deal of thought typically goes into such a visually striking part of the home. If a client has contracted for the home, she will most likely choose the cabinets. If this is a spec home, the project manager may be tasked with the job. It's usually best for the project manager to rely on the advice of an interior designer or the designer associated with the cabinet company.

There are basically two options to consider when selecting a source for the cabinets: custom or prefabricated. Prefabbed cabinets can be purchased from any major building supply company in a variety of styles and finishes; some even allow for minor customization. However, there is nothing quite like having a company build a custom set of cabinets for the home. The options for customization are vast. By visiting the job site to measure and design, the decorator can ensure a perfect fit. Custom cabinet shops can also be surprisingly affordable.

There is also the question of countertops. Often the cabinet company provides the countertops, but this is not always the case. A different company may provide and install the countertops. If this is the case, after the cabinet subcontractor measures and creates a cabinet design, this design will need to be sent for the countertop subcontractor to begin work on the various countertop needs.

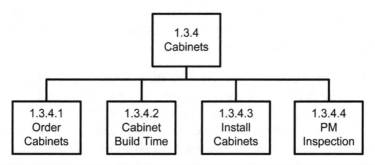

Figure 4.14 WBS Section 1.3.4

Whether custom or prefabricated, the cabinets should be ordered after the framing is complete. This allows the provider to come in and measure the spaces to ensure the right sizes are ordered. She will typically mark on the floor where the various fixtures will go, as well as provide a schematic to the project manager so that the placement of all the fixtures and lighting can be verified with the electrician, plumber, and HVAC subcontractor.

1.3.4.2—Build Cabinets

After the cabinets have been ordered, the project manager must allow for time to either build the cabinets or have them shipped to the job site. Ordering the cabinets after the framing is complete typically allows plenty of time for the cabinets to be built. If a custom shop is used to build the cabinets, they will typically hold the cabinets at the cabinet maker's warehouse until they are needed at the job site, which is a great benefit to the project manager.

1.3.4.3—Install Cabinets

Before the cabinets are installed, the project manager should inspect the subflooring one last time to ensure that there are no issues. Depending on the number of cabinets in the home and the size of the installation crew, the job can take from one to a few days to complete. The crew typically installs the base and overhead cabinets, and then moves to installing the doors, shelves, and trim.

Next the countertops are installed. If the project calls for a solid-surface countertop, such as granite, the countertop subcontractor will often install the kitchen sink as well. If she does not offer the service, the sink will need to be discussed with the plumbing subcontractor to make sure the correct look is achieved.

1.3.4.4—Inspect Cabinets

The project manager should inspect the workmanship to ensure that all doors, drawers, and shelves are well balanced and that they close properly. The project manager should take steps to protect the cabinets throughout the remainder of the construction project. The cabinets are one of the few solid surfaces on the construction site, so people will use them to mix paint, repair tools, eat lunch, and so on. To avoid damage to the cabinets, they should be covered, and workers should be instructed not to use them as a work bench.

1.3.5—Paint When the time comes for the home to be painted, the client is typically very excited, as paint starts to give the home a finished look. The project plan will specify the various types of paints to use, the different types of application, and the colors to be used. There are as many pitfalls at this point of the project as there are possibilities. The paint will reveal whether a high-quality job has been done on the walls and ceilings, as it often shows every blemish. The steps associated with this section are (see Figure 4.15):

1.3.5.1—Order exterior painting materials
1.3.5.2—Paint exterior of home
1.3.5.3—Order interior paint materials
1.3.5.4—Paint interior of home
1.3.5.5—Inspect paint

1.3.5.1—Order Exterior Painting Materials

Although this section focuses on the interior finishes, the exterior work is mentioned here for convenience. Some may decide to create a new section below the Exterior Work & Finishes, which would be fine. But for the sake of this example, it will be dealt with here. Depending on the type of materials used on the exterior of the home, there may actually be very little to do. Vinyl siding, brick and rock veneers, and other such surfaces do not need to be painted. Many times, the trim on the exterior of the home, such as soffits and eaves, are also vinyl.

Regardless of the amount of surface to be painted, the project manager must ensure that the products ordered and delivered are exactly what have been requested by the client for this project. Paint involves a number of variables, such as type, brand, color, and the like. Now it might turn out that the clients are unhappy with the color when it is actually applied to the home, but the project manager's job is to make certain that she is in compliance with the project plan.

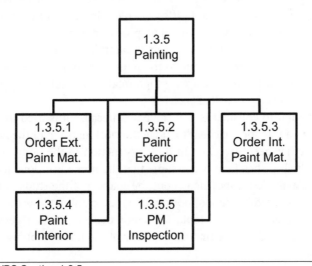

Figure 4.15 WBS Section 1.3.5

1.3.5.2—Paint Exterior of Home

Painting the exterior of the home is a relatively straightforward phase. The project manager first ensures that the surfaces are adequately prepared before the paint is applied. This task includes cleaning the surface and doing the necessary prep work, such as caulking and sealing joints. As part of the planning for this phase, the project manager discussed this with the painter, so that he understands what is expected. A wise project manager will do a brief inspection after the surfaces have been prepared to make certain everything is as it should be. Many times, painters will fill small cracks in joints with paint rather than caulk as they should be. Once the paint is applied, no one is the wiser; but after a little exposure to the elements, the paint will pull apart and the crack will be visible.

The project manager should do a thorough inspection after the painting is complete. He should use a ladder to inspect those areas that are less visible from the ground. Some painters have been guilty of only applying a thin or single coat in the higher eaves of a home, so the project manager must verify that the work has been properly completed. It should also be noted that as the project progresses, it is not uncommon for someone to damage the finish by leaning a ladder against the home or inadvertently hitting it with a piece of lumber. Because of this, it is a good idea to include in the contract a return trip from the painter to touch-up these areas before the project is completed.

1.3.5.3—Order Interior Paint Materials

In the interior of the home, the steps and guidelines are roughly the same as for the exterior. The primary difference is that in the interior, the work is typically more complicated. The clients will want different colors in different rooms with different finishes and in a variety of styles. The project manager must carefully select the painting subcontractor to be sure this is carefully handled. Specialty finishes are much more in demand now due to the popularity of home renovation television shows, which often feature a very high quality of workmanship. This is the standard that some clients will expect.

Another point to consider is that the paint, because it is used throughout the house, provides a way for the client to judge the workmanship of the home. If great detail and care has been exercised in the construction of the home, but the paint job is of low quality, the client will assume that the home is of low quality overall. The client is usually less concerned about the spacing of the studs in the wall and the grade of wood used in the subflooring than they are about how the cabinets, the finished flooring, and the paint look. Therefore, the project manager's investment of extra effort in making sure that the painting of the home is of high quality will more than pay off.

1.3.5.4—*Paint Interior of Home*

After the project manager has verified that the right paint has been delivered, work is ready to begin. As most job sites are not climate controlled, the project manager must ensure that the painter is able to work under the best conditions possible. She must also verify that the painter is prepping all the surfaces as agreed. Smooth joints, clean walls, and caulked nail holes are imperative for a quality paint job. As the painter progresses through the home, the project manager should inspect the work in such a way as not to be a nuisance, but to make certain that the painter understands he will be held accountable for the work. If the project manager notices a problem early on, it should be addressed early on. If she took the time in the planning phase to select a reliable painter, there should be few problems.

In many cases the same paint subcontractor will do the interior and the exterior work; if not, the work contract with the painting subcontractor must include a clause whereby he is obliged to return to the job to touch-up the interior paint work after the home is complete. The darker the colors and the more intricate the paint style, the more difficult it is to touch up the paint. A small blemish may require an entire wall to be repainted. The client should be warned that the selection of certain paints will make it very difficult for them to be color matched in the future.

1.3.5.5—*Inspect Paint*

After the work is done, the project manager needs to complete a thorough inspection of the home to make certain that the work was completed according to the agreed upon standard—are the colors correct, finishes correctly applied, etc? If not, it will need to be addressed, and the work corrected.

1.3.6—Flooring Once the flooring is installed, the home really begins to have a move-in feel to it. This part of the WBS only contains a few steps, but these steps may turn into a few different entries on the actual project schedule, as the flooring may be installed in more stages than are shown in Figure 4.16. Installing flooring in the home can be a challenging process, but if a clear plan has been developed and is faithfully executed, the process proceeds smoothly. The steps are:

1.3.6.1—Order flooring materials
1.3.6.2—Install flooring in kitchen and baths
1.3.6.3—Install flooring in remainder of home

1.3.6.1—*Order Flooring Materials*

As the client or the project manager selects the flooring materials in the planning phase. Included within that plan should also be the appropriate lead time for

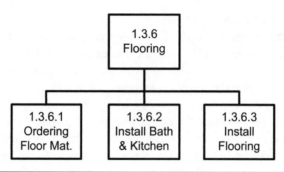

Figure 4.16 WBS Section 1.3.6

obtaining each material type. Many times, it is necessary for the project manager to use multiple sources for materials. Tile may come from one supplier, hardwoods from another, and carpet from a third. The project manager also has to take into account storage of the materials. Many times, the vendor will hold the materials until the project manager needs them delivered to the job site, which is a great advantage if available. Whatever the case, the project manager ensures that the materials are ordered in a timely manner so they can be on the job site when the work is scheduled to begin.

1.3.6.2—Install Flooring in Kitchen and Baths

The flooring in the kitchen and bathrooms is often installed before the floors in other areas of the home. The reason for this is that the flooring must be in before the kitchen and bathroom fixtures are set by the plumber. If vinyl is being used, the process goes very quickly; but if a material such as tile is specified, the process may take a few days.

The project manager should take the time to inspect the surfaces with the flooring subcontractor to make certain that no issues will arise once the work begins. If a floor is uneven, it is best to address it before the tiles are set in place. As the work progresses, the project manager alerts other workers not to disturb the area until the flooring subcontractor authorizes traffic over the area. After the flooring is installed, the project manager should inspect the work to make certain that it has been completed in compliance with the specifications of the project plan.

Kitchen and bathroom floors also need to be protected, typically by covering the surface with a durable paper that can withstand the expected foot traffic.

1.3.6.3—Install Flooring in Remainder of Home

The second phase of work is to install the flooring in the remainder of the home, according to the construction schedule. The timing of the installation will depend on the type of flooring being installed. If carpet is being installed, the process will most likely take a few days. Appropriate steps should be taken after it is installed to create walkways across so that it will not be soiled by work boots. If, however, hardwoods are being installed, there is more to consider, depending on the type of wood being used. There are basically three choices: (1) an engineered hardwood, such as laminate, (2) a prefinished hardwood, or (3) an unfinished hardwood. Each of these has its own strengths and weaknesses, which are beyond the scope of this book, but some basic guidelines are possible.

The least time consuming are the engineered hardwoods and the prefinished. They can simply be installed according to the manufacturer's instructions. The most time consuming to install are the unfinished hardwoods, which require sanding and finishing after installation. This can create a delay in the project schedule if it has not been adequately planned. The process requires that the areas be vacant until the flooring subcontractor completes his work and authorizes reentry into those areas, which can take a few days, depending on the size of the job. As the work progresses the project manager should perform the standard type of review inspections to see how the work is progressing. If something seems to have been done unsatisfactorily, it is best to address it before the finish has been applied.

After the work is completed and reentry is allowed, the project manager should inspect the floors thoroughly to make certain that the joints are tight and that the finish has been applied evenly to the surface of the floors. The project manager should take the appropriate steps to protect the floors for the remainder of the construction project.

1.3.7—Finish HVAC, Plumbing, Electrical, and Specialty Rough-ins This is a rather broad category in the WBS, which encompasses the work required to complete the various major systems of the home, such as electrical, plumbing, heating and cooling, as well as any others that the project plan calls for (see Figure 4.17). The work that is associated with each of these systems will in some form integrate with the other systems, so this must be planned and coordinated. An example of this interaction is that the hot water heater can act as a backup for the heating system if the HVAC contractor installs the necessary materials to accomplish this. Or the HVAC subcontractor can install a heat pump, but the electrician must connect it to the electrical system of the home. In this way, each subcontractor may come and work but then must return later to complete the job. The project manager must closely monitor the progression of work and adherence to the schedule. The subsections of the WBS associated with this phase are:

1.3.7.1—Finish HVAC
1.3.7.2—Finish plumbing

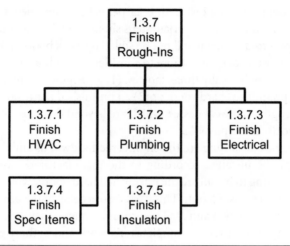

Figure 4.17 WBS Section 1.3.7

1.3.7.3—Finish electrical

1.3.7.4—Finish specialty items

1.3.7.5—Finish insulation

1.3.7.1—Finish HVAC

Finishing the HVAC system of the home typically includes the installation of any air lines not previously installed and the installation of the major mechanical systems, such as the heat pump, air handler, gas furnace, for example. Depending on the size of the home, this work may require a few days to complete. Because there is limited power available through the temporary power pole, the ability to test many of these systems thoroughly is limited. However, this is common, and the systems will be thoroughly tested once the building inspector's office authorizes power to be turned on for the final inspections. Depending on the project manager's knowledge of these types of systems, her ability to inspect work prior to that point might be rather limited. But she should take steps to inspect all that can be visually inspected.

1.3.7.2—Finish Plumbing

Next in the list is the plumbing. At this point, the plumber has already run all the lines. Now the focus is installing sinks, faucets, the water heating system, and the like. The ability to test these items is also rather limited, unless the water is connected to the home and the project manager can gain authorization to turn it on from the local water supplier. The project manager can, however, visually inspect the work to ensure that the correct faucets, sinks, and water heater are installed. Such an inspection can save the project manager some trouble. For instance, the client notices that the incorrect faucet was installed. There is nothing that breeds distrust in a client faster than the project manager's missing something that is so obvious to the client.

1.3.7.3—Finish Electrical
The electrical work done by the electrical subcontractor is also difficult to inspect until power is supplied to the home. However, the project manager can inspect the more visible portions of the work. He should ensure that the proper electrical outlet covers are used, if the client has requested a specific type. He can verify that the correct light fixtures have been installed and that they are properly balanced. The true inspection will come once the power is turned on in the home.

1.3.7.4—Finish Specialty Items
This section of the WBS is rather broad. The major systems have been dealt with, but there are other systems that the client might have elected to install, which might require a return visit at this point in construction. Some of the systems might include an alarm system, a central vacuuming system, an audio/visual system for an entertainment room, or a host of other features that are available. Whatever the system is, the project manager is responsible to make certain that the work is completed according to the specification provided in the project plan. Adequate time and care must be taken to ensure that these items are completed.

1.3.7.5—Finish Insulation
The last item in this section is insulation. Previously, the floors and walls were insulated according to the project plan; this step involves insulating the ceiling. There are a number of types of insulation that can be used; the most common in my area is the type of insulation that is blown. Depending on the size of the home, this should not take more than a day or so to be completed. The project manager can visually inspect this work by going into the attic area and verifying proper coverage and thickness.

1.3.8—Trim Finishes This section refers to a number of small tasks that must be done throughout the home toward the end of construction. In the sample WBS (see Figure 4.18), only three tasks have been included, but there may be others depending on the particular needs of the project. In the example, these items are listed:

1.3.8.1—Install door hardware
1.3.8.2—Install bathroom hardware and accessories
1.3.8.3—Project manager inspection

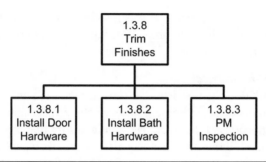

Figure 4.18 WBS Section 1.3.8

1.3.8.1—Install Door Hardware

Installing the door hardware throughout the home is not limited to merely installing door knobs. It may also include adjusting doors so that they open and close properly, installing door stoppers, and even changing hinges, if necessary. The one who performs this work should be instructed to do it carefully. A sloppy job will require rework, and it may lead to having to repaint the door.

The project plan should include the type of door hardware to be installed, but if it does not, the project manager should consult with the client. If the wrong style or finish is installed, it can be quite time consuming to go through and change out all the handles. Because the price and quality of door knobs varies, the client may choose a fancier knob that is made of lower quality because she will get the look she wants for the price she can afford. The project manager should warn her about the issues that can arise from selecting a lower-grade handle. The project manager should certainly encourage that a quality knob and deadbolt be purchased for exterior doors. Depending on the size of the home, this task will take from a day to a day and a half on most projects.

1.3.8.2—Install Bathroom Hardware and Accessories

Once again, this is a broad category that may need to be customized to fit the particular project. Time must be set aside to take care of these mundane tasks. Hanging towel racks, shower doors, toilet paper holders, mirrors, medicine cabinets and other such items should be handled by an experienced individual. These items draw the eyes of the client; he wants to make certain that they have been placed appropriately on a wall and that they are level. If a he sees a towel rack that has been hung out of level or a shower door that leaks, he will question the project manager's competence. In his mind, this is such a small thing; if the project manager hasn't done properly, what else is done wrong?

Many times, the specific item to be used will be listed in the project plan. If not, the client should be consulted. Most buyers are rather particular about these detail items. The project manager might even provide the client with a list of needed items and a budget, and ask him to select the items.

1.3.8.3—Project Manager Inspection

After the door hardware and the bathroom hardware and accessories have been installed, the project manager should perform an inspection. He should make certain that all the door knobs are tightly installed, that the door stoppers are properly placed, and that the doors hang evenly and are well-balanced. He should also make certain that all the bathroom hardware and accessories have been securely fastened to the wall so that they will withstand the weight of having clothing and wet towels draped over them.

Section 1.4—Exterior Work and Finishes

It is important to recall that the order of the WBS does not necessarily reflect the actual order in which the work is completed. For instance, the first step listed in this section is the final rough grade. According to the WBS, it would be done after the trim finishes, but this is not the case. According to the project schedule, the exterior work could be completed much earlier. The project manager should reference the project schedule, not the WBS, to see the actual order in which the work should progress.

The exterior work and finishes section (see Figure 4.19) includes some of the more exciting parts of the construction project. But it is also one of the most frustrating parts of the project, especially once the final inspections and walk-throughs begin. Regardless of whether the project is behind schedule, the client will most

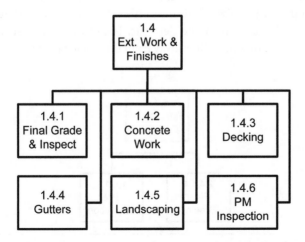

Figure 4.19 WBS Section 1.4

often act as though it is, once she sees that the end is in sight. She will begin asking when she can begin moving furniture into the home and other such questions. The project manager needs to hold firm to the project schedule. The project schedule should reflect the most realistic timetable for completing the project. The project manager should not be pressured into making unrealistic promises about how much sooner the project can be completed. Keeping the client updated as to the progression of the project will help manage her expectations.

The exterior work and finishes section includes:

1.4.1—Final rough grade and inspection
1.4.2—Concrete work
1.4.3—Decking
1.4.4—Gutters
1.4.5—Landscaping
1.4.6—Project manager inspection

1.4.1—Final Rough Grade and Inspection From the time that the lot is pre-pared for construction to the time that the drywall is delivered, a number of trucks and other heavy equipment have been over various parts of the construction site. During the initial site preparation, much time might have been invested in clear-ing all the necessary portions of the lot—or the minimal amount of work to begin construction might have been done. If so, now is the time to complete any clearing and rough grading (see Figure 4.20).

The goal of the rough grading is to prepare the sight for the driveway, patios, decks, and landscaping. The person who does this work must have a strong grasp of how to ensure that the land will be able to manage the normal amount of rainfall and snow that the area receives. This means that the land must be shaped in such a way as to protect the home from the elements. The project manager should ensure that the water will be turned to flow away from the home on all sides, so that water

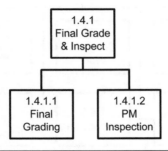

Figure 4.20 WBS Section 1.4.1

will not flow up against the home and soak through the foundation wall over time. It also requires that there be no low points in the yard where the water will stand after a rainfall. Many grading subcontractors either lack the skill or the knowledge to do this type of work; the project manager must locate a skilled finish grader to perform the work.

Once the grading has been done, the project manager should inspect the work to make certain that all is as it should be. If there are small areas to be addressed, the project manager should have those areas corrected before payment is made.

1.4.2—Concrete Work This section refers to driveways and patios that the project plan specifies. It is helpful to think of this section in two parts (see Figure 4.21):

1.4.2.1—Grade and Form Driveway and Patios
1.4.2.2—Pour Driveways and Patios

1.4.2.1—Grade and Form Driveway and Patios
It is helpful to think of this in two steps because it will break up the work so that the project manager and client can inspect how the concrete will look prior to its actually being poured, at which point modifications become decidedly more difficult. There is not much grading to be done after the rough grading is complete, but the concrete subcontractor typically grades the areas for the concrete and patios, which may include hauling in dirt to the job site.

After the concrete subcontractor has graded the area, he will form up the areas where the concrete will be poured. This will allow the project manager and the client to clearly visualize the areas where the concrete will actually be poured. Any modifications should be made at this point after the client has reviewed the

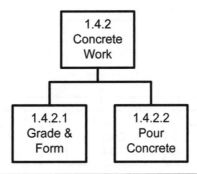

Figure 4.21 WBS Section 1.4.2

forms. It is usually a good idea for the project manager to meet the client with the concrete subcontractor so that he understands any changes that the client is requesting. If changes are made, the project manager determines how this might affect the cost and schedule of the project. Once the cost is determined, the project manager executes a change order and collects the extra money according to the terms of the contract. Implementing any changes prior to performing a signed change order should never be done. The client has the right to know the cost and approve it before the work is done.

1.4.2.2—Pour Driveways and Patios

The project manager should take steps necessary to protect the concrete after it has been poured. The concrete subcontractor should provide information as to how long the concrete needs to cure before it can be walked on. The project manager must ensure that the concrete is not damaged by the various workers who will follow after the concrete has been poured. This may mean erecting a permanent barrier so that workers do not drive on it. There are few events more annoying to the clients than having a subcontractor's '84 Chevy truck leaking oil on their new driveway. An inspection of the concrete work concludes this step.

1.4.3—Decking The next item listed in the WBS is the exterior decking (see Figure 4.22). Depending on the size and style of the home, some of the decking may be completed much earlier in the project. For instance, if the plan calls for a covered porch, it would most likely be roughed in and completed when the home is framed. However, other deck work might still require completion in this phase.

Depending on the project plan, multiple decks may need to be built or no decking may be needed. The cost of decking is typically a surprise to most buyers. Even a basic wood deck with simple materials may seem rather expensive to

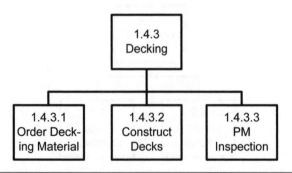

Figure 4.22 WBS Section 1.4.3

some. However, the pleasure that a well-planned deck can provide to the client often makes it worth the cost. The project manager must ensure that the clients are aware of cost and approve it.

It is important to choose the right individual to construct the decks. Often, the trim carpenter performs the work, sometimes another subcontractor is engaged. The selection should be based on skill and proven craftsmanship, not an inexpensive bid.

If the deck is somewhat unique in any way, the project manager must make certain that the subcontractor has a clear understanding of what the client desires. Detailed drawings or pictures can be very helpful.

As the work progresses, the project manager might drop by periodically to see that the work is progressing according to plan. This way, he ensures that if there is a misunderstanding, it is addressed early on. After the work is completed, the deck should be inspected and payment authorized.

1.4.4—Gutters The gutters are a seemingly insignificant part of the construction project, but they play an important role. This section contains two entries on the WBS, which can be seen in Figure 4.23. Poorly designed and installed gutters and drain lines are one of the primary causes of water in the basement or crawl space. Today, there are a variety of styles and colors from which the client can choose. Some clients even opt to not have gutters installed on the home for aesthetic reasons. If this is the case, the project manager should warn that water issues might develop, and she must take steps to mitigate those issues. After the gutters have been installed, the project manager should review the work.

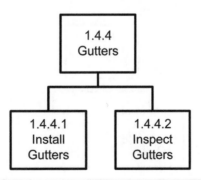

Figure 4.23 WBS Section 1.4.4

1.4.5—Landscaping Depending on the size and nature of the project, landscaping may include as little as sowing grass on the disturbed areas. However, if the client's budget allows, it may include much more. Most likely, the project manager will simply oversee the work done by a subcontractor. If this is the case, the project manager's concern is two-fold (see Figure 4.24). First, the project manager takes into account any guidelines provided by local zoning officials. For instance, in some municipalities, the contractor is responsible for planting a certain number of trees of a certain size. If there are such guidelines, the project manager must make certain that the landscaping subcontractor includes those required items as part of the contract. Second, the project manager should ensure that the landscaping subcontractor performs the work according to the design and specifications outlined in the project plan. This can be done by reviewing the work as it is completed and then by performing a final inspection before payment is authorized.

1.4.6—Project Manager Inspection The last step of this phase is an inspection by the project manager of all the work that has been completed as part of this section of the WBS. Inspections have been done as the work progresses, but there is still benefit in performing a final walk-through to ensure that the work has been properly completed.

Section 1.5—Final Inspections

Now the project manager has finally reached the last section of the WBS—final inspections (see Figure 4.25). This can be both a rewarding and a stressful time, as the project manager is combing through the project ensuring that all is ready

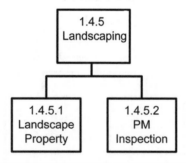

Figure 4.24 WBS Section 1.4.5

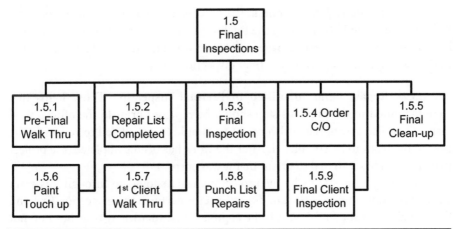

Figure 4.25 WBS Section 1.5

for the inspections. Depending on the local guidelines, the final inspection may be a one- or two-step process, as outlined in the WBS. The steps included in this WBS are:

1.5.1—Prefinal inspection walk-through
1.5.2—Repair items identified in prefinal inspection
1.5.3—Final inspection
1.5.4—Order certificate of occupancy
1.5.5—Order interior and exterior clean-up
1.5.6—Paint touch-up
1.5.7—First walk-through with client
1.5.8—Punch list repairs
1.5.9—Final walk-through with client

1.5.1—Prefinal Inspection The prefinal inspection is conducted by the project manager and any other team members who possess the necessary skills. A checklist is the best way to conduct the inspection. A checklist can be developed by consulting with the local building inspector about what she checks on the final inspection. Although the project manager may not possess the expertise necessary to inspect everything as thoroughly as the building inspector can, he should be able to perform a fairly thorough inspection aided by the checklist. As the inspection is performed, the project manager creates a list of any items that need to be repaired. The list should also include any items that he is uncertain about. It is important to note that the project manager should take the time to note cosmetic issues, such as a scuffed wall or something similar, but those items need not be repaired until a step a little later on in the process. If the project manager orders a wall to be repainted, then during the final inspection some problem is discovered behind the wall, the project manager will have wasted the time and money on having the wall repainted prior to the final inspection.

1.5.2—Repair Items Identified in Prefinal Inspection Those items that were identified in the previous step should be repaired prior to the official final inspection. The appropriate subcontractors should be scheduled to return to the site for the work to be completed. After the work has been completed, the project manager reviews it to make certain that it was corrected.

Some might question why the project manager should go through this extra effort. Why not simply let the building inspector perform the inspection first and then repair what she finds? This is the approach that many may take, but I do not advocate this for two reasons. The first is a pragmatic reason. Whenever a portion of the project fails an inspection, it reflects poorly on the project manager. It can appear to the client that the project manager is either incompetent or ignorant, neither of which is desirable.

Second, the project manager who opts to wait for the building inspector to pass judgment appears to lack pride in the work. He is merely satisfied to get by, to meet the minimum standard. The project manager should strive for excellence in the work; he should have pride in his abilities and skill in creating a well-built home. He should take the time to inspect the work and make certain that it is done properly for the sake of his reputation, not merely because the building code regulates the work. He should self-regulate the work, holding it to a high standard.

1.5.3—Final Inspection After the items identified in the preinspection have been repaired, the project manager can schedule the final inspection. The project manager should plan to be onsite for the final inspection. Each locale may have particular requirements for the final inspection with respect to electricity, gas, and water. The project manager should review these requirements with the building inspector's office to make certain that all is ready for the final inspection.

As the building inspector performs the final inspection, the wise project manager follows along, making notes about the inspection. If the inspector notes something that needs repairing, the project manager can make any notes that will be helpful in instructing the subcontractor. Or the project manager might learn something that she did not previously know about the inspection, which she can jot down to plan for on future projects. In this way, the project manager is using this step as a lesson that can be applied to future projects.

After the inspection is complete, any repairs should be made as soon as possible and the reinspection passed. It is my opinion that the client should receive full disclosure regarding the inspection and should be informed of the steps taken to correct any issues. In this way, the project manager can manage any negative responses that might arise immediately. Once the repairs have been made, the reinspection is ordered. The project manager should, once again, meet the inspector on the site to review the items. After the home has passed inspection, the project manager should update the project plan with any special notes which may be helpful for future projects.

1.5.4—Order Certificate of Occupancy After the home has passed the final inspection, the project manager orders a copy of the Certificate of Occupancy. This certifies that the home has in fact passed the final inspection, which simply means that it has the approval of the governing body to be occupied by a resident. Typically, the building inspection office forwards a copy to the power and water companies so that the client will be able to turn the power and water on in her own name. But the project manager needs a copy for his files, a copy to give the client, and a copy to forward to the bank and closing attorney.

1.5.5—Order Interior and Exterior Cleanup After the home has passed final inspection, it is cleaned for the final walk-through. As construction progresses, the trash and material waste is loaded into either a dumpster or truck and hauled off. This might be needed now, but the cleanup also includes a thorough cleaning of the interior of the home by a professional cleaning service. It should include cleaning the following items:

- Walls
- Baseboards
- Floors
- Bathrooms
- Windows
- Light fixtures
- Cabinets
- Basement
- Driveway
- Exterior

The cleaning should be thorough enough that the client can move into the home without doing any additional cleaning. After the cleaning crew is complete, traffic to the job site should be kept to a minimum.

1.5.6—Paint Touchup Once the home has been cleaned, the project manager should inspect the home to see if there are any areas where the paint needs to be touched up. Most likely, there will be a few scuffs on the walls or on doors.

Touching up the paint can be a tricky job. If the client has chosen a dark or deep color, depending on the size of the blemish, the entire wall might have to be repainted. It is often very difficult to touch up one portion of a wall if a dark color has been used or a special finishing technique has been employed. The project manager will need to rely on the expert opinion of the painter to determine what needs to be done for the blemish to be properly removed.

1.5.7—First Walk-through with Client After the home has been cleaned and the paint has been touched up, the project manager is ready for the first walk-through with the client. Here again a checklist helps the project manager guide the client through the property, pointing out how the finished product matches her desires and expectations. By leading the inspection in this way, the project manager ensures that most of the items identified by the client will be caught in this first inspection, so that she is not continually adding items to the list.

Once the inspection is complete, the project manager should go through the list one more time with the client to make certain that both parties have a clear understanding of what the issues are and how the project manager will address them.

The project manager should be glad to address most items that are on the list, but conflict can arise. If the client believes that a portion of the project is not in compliance and should be corrected, but the project manager believes that the work has been completed properly, how are such conflicts resolved? First, the project manager ensures that he clearly understands the exact nature of the client's complaint. He needs to truly understand the issue before he can begin to effectively manage the complaint. After hearing the complaint, he has a few options. First, if the complaint is valid, he must correct the problem. If he believes the complaint is invalid because the buyer is misunderstanding something, he can attempt to explain it in such a way as to clear up the confusion. If, however, the client and the project manager reach what appears to be an impasse, one of the two must bend or the issue will be headed for either arbitration or court, which should be avoided if possible. If the project manager believes he will not be able to resolve the situation, he should contact his immediate superior. It is important that during such a situation, he document all conversations related to the issue so that he can pass this information along.

1.5.8—Punch List Repairs The list of repairs to be made following the final inspection is called a *punch list*. After the punch list has been finalized, the project manager sees that all the work is completed as soon as possible. Care should be taken to make certain that the job site remains as clean as possible so that a future cleaning will not be necessary. As the workers complete their repair work, the project manager should inspect the work and approve it.

1.5.9—Final Walk-through with Client In the final walk-through, all the issues that were previously identified should be addressed to show that they have been remedied. Once the client approves the repair work, it is a good idea to have him sign an approval form accepting the work.

Concluding Comments on the WBS

The primary focus of the execution phase of the project is doing the actual construction work outlined in the WBS. But there is other work during the execution phase of the project, as there are other parts to the project plan that must be considered in addition to the WBS. This will be the focus of the remainder of the chapter.

Executing Management Plans

The WBS has been the major focus of this chapter, but it is one part of the project plan. Without question, it is the most visible, but there were other plans developed. These include:

- Scope management
- Cost control
- Quality control
- Human resource
- Communication
- Risk management
- Purchasing and contract administration

These other plans also require work during the execution phase of the project. They were not prepared for the sake of practice during the planning phase—they are to be used. Large portions of each of these plans will be used during the controlling phase, which is covered in the next chapter of the book. Therefore, only a brief overview of the work associated with each section during the execution phase of the project is discussed here.

Scope Management

The scope management plan contains the boundaries of the project, as well as instruction as to how those boundaries can be changed. The plan for monitoring those boundaries is covered in the controlling phase, which is the next chapter, but there will be work associated with this plan during the execution phase. If during the construction project, the client were to modify the project in some way, the change would be managed through the integrated change control system, which is covered in the next chapter. However, once the plan has been changed to include the work, the new work or rework must be implemented, and this work will be considered part of the execution phase. Therefore, during the execution phase, once the change orders are made, the changes are implemented, which means that the work schedule will need to be adjusted. In this way, the scope management plan will in some ways impact the execution phase of the project.

Cost Control

During the planning phase cost estimates are prepared and budgets created based on estimates. During the execution phase, the materials are ordered, and the subcontractors are paid. On smaller projects, the project manager is likely responsible for writing and mailing the checks and keeping records; on larger jobs, there is another team member tasked with handling the payment process.

During the planning phase, the cost control plan is created, stating how this process should be done. And during the execution phase, the process must be followed. The person responsible for managing the accounts payable of the project executes the various directives of the cost control plan throughout the execution of the project. This typically includes processing orders and paperwork, reviewing invoices and bills, and making payments according to the terms of the contracts. The work done by this person will be monitored by the project manager, but, in general, this person is responsible for ensuring that all this work is done according to the plan developed. Therefore, the project manager is responsible not only to make certain that the actual construction work is being done properly, but also for ordering the work outlined in the cost control plan making certain that all parts of the project are executed properly.

Quality Control

The quality control plan involves two distinct aspects as outlined in the previous chapter. First, quality refers to the way that the project is managed. Thus, the project follows the project plan. This task is primarily handled in the controlling phase of the project. The second aspect is quality assurance in relation to the product being built, and this is considered part of the execution phase of the project. As a matter of fact, inspections are so important that they are inserted into the project schedule at various points and in the WBS. Because of this, information related to inspections was primarily covered in the first part of this chapter. But it is important to note that these inspections should be given priority during the execution of the project.

Human Resource

The human resource plan contains the overall plan for who will perform the various tasks of the project. This includes not only those who actually work on the building, but also those who perform behind the scenes managing the project. During the execution phase, the project manager typically just implements the plan that has been developed. But there are times when a subcontractor is no longer able to perform the work or when a team member is assigned to a different project. In these instances, the project manager is responsible for finding a new person to fill the gap. If this is not done promptly, the project suffers due to time delays or increased costs—even the quality of the project can suffer. The project manager must keep this plan up to date and must make certain that the plan developed before the actual construction began will remain valid as construction continues.

One of the biggest challenges that the project manager faces during the execution of the project is managing the various personalities of the people working on the project. As in any job setting, construction personalities can clash, and the project manager must be able to handle friction so that it does not create a negative impact on the project. Therefore, during the execution phase of the project, the project manager must be aware of the types of issues that can arise and be ready to deal with them. Most of the time, conflicts arise due to people misunderstanding their role. The project manager can avoid most problems through clear and open communication. This will not eliminate all interpersonal conflict, but it will make it much more manageable.

Communication

The communication plan is one of the most critical plans during the execution phase of the project. Most problems arise due to communication failures of a few different types. One person says something to another who hears the words the sender is saying, fails to understand the meaning, but believes he has a clear understanding of the intended message. The person then takes this information, which he has not understood, and uses it to either perform his work or instruct another regarding her work. The result is that the work is not performed correctly, which can lead to cost increases and time delays.

The best way to avoid this problem is to have as much direct communication as possible with those actually performing the work. The fewer people the information passes through, the closer it will resemble the original message. A second way to avoid miscommunication is to have people repeat back what they understood. A final suggestion for avoiding miscommunication is to use drawings whenever applicable. Typically, if someone can see what she is being asked to do, she will be more likely to understand it. These are but a few of the methods that can be used.

Many times, information is communicated correctly, but it arrives too late. This is why developing a good communication plan includes not only the information that needs to be sent to a person, but also the time when the person must have the information. This is even more critical in construction projects, where many aspects are constantly being modified and changed due to the dynamic nature of the work. The project manager must be vigilant in making certain that the subcontractors, team members, and vendors are receiving information at the appropriate time.

With these common problems in mind, the project manager follows and updates the communication plan during the execution phase of the project. As the dynamics of the project change, the plan will need to be updated to reflect any necessary modifications. If the project manager does not do this, the risk of making mistakes is increased.

Risk Management

The risk management plan is primarily associated with the controlling phase of the project. This is because the plan contains a list of possible events that could either harm or help the project, and the project manager must watch to see if any event occurs that creates the need to modify the project in some manner to compensate for the risk event. When a risk event occurs, the project manager executes the plan as part of the execution phase of the project.

Once a risk response plan is determined to be necessary, the project manager assigns the staff and resources necessary to implement the plan. This may require modifying the schedule for the overall project to accommodate these unexpected developments. Once the solutions have been implemented, the project manager shifts resources back to the planned tasks to continue the planned work of the project.

Purchasing and Contract Administration

A vital part of the execution phase is implementing the purchasing and contract administration plan. This plan contains the guidelines and details regarding purchasing the various materials to be used in constructing the home. Implementing this plan requires more than simply ordering materials on time; it also includes inspecting orders once they arrive and managing the materials on the job site.

Managing materials on the job site is critical to any project. Sometimes materials have a way of just simply disappearing. This can occur from someone making a mistake and moving the materials to a different job site, or it can occur because of theft. A few years back, the company I work for had an order of vinyl siding go missing from the job site the night before it was to be installed. The subcontractor called us looking for the material, which he thought had not been delivered. After going to the job site to inspect, we decided to drive through some of the other active neighborhoods nearby just in case. It was a long shot, but because the vendor always wrote our company's name on the boxes delivered to the job site, we thought we would take a chance. In the second neighborhood we drove through, we found the materials being installed on a house and half the boxes still unopened and sitting in the yard . Needless to say this was an open-and-close case for the local police to investigate. It seemed that the guy who stole the material thought he could get it up and throw the boxes away before he was discovered.

Another aspect that must be managed properly is what to do with extra materials. If there are extra materials on the job site after the work is completed, they must be returned or moved to the next job site. In either case, the job they were removed from is credited to reflect the fact that not all the materials were used.

The second aspect of this plan involves managing the contracts entered into with the various subcontractors and vendors, as well as the contract between the construction company and the client. Someone must be assigned to keep up with these documents and to make certain that the work performed is in agreement with the work contracted. If the project manager does not have these contracts reviewed and moderated, the construction company is responsible for correcting any problems that arise because they were negligent in their responsibilities as it relates to the management of the contracts. At various points, the contracts should be reviewed to ensure that all parties are in compliance with the terms of the agreement.

TRANSITIONING TO THE CONTROLLING PHASE

Not only must the project be executed, but as the work progresses it should be monitored so that deviations from the project plan may be corrected when they occur. Therefore, the next chapter will cover the fourth phase of the project lifecycle: controlling the construction project.

5

CONTROLLING THE CONSTRUCTION PROJECT

So far, the first three phases of the project lifecycle have been covered:

1. Initiating the construction project
2. Planning the construction project
3. Executing the construction project

These phases progress in a fairly linear manner. Once the project is initiated, then it is planned. Once it has been planned, the time arrives to execute the plan. While the execution phase is in progress, there is another phase also in process—the controlling phase.

The controlling phase of the project lifecycle contains processes that are necessary to accomplish two tasks. First, the processes ensure that the plan is being followed in the execution phase of the project. To accomplish this, the project manager uses a number of performance evaluation techniques and inspections, which compare the actual work of the project to the planned work of the project. The comparison provides data for informed decisions about the execution of the project. In this way, the data are used to correct any problems or deviations that arise during the execution phase of the project.

The second focus of the controlling phase is integrated change control. When the project plan was developed, it was done so with the knowledge that a construction project is not performed in a vacuum. Situations arise that will require modifications to the project plan. For instance, the client might change her mind; the project manager might have to adjust the plan because of poor planning; or unforeseen events occur. Regardless of the reason, any experienced project manager will recognize that changes are inevitable and makes plans so that any disruption will be minimal. She develops an integrated change control system to be used as the need for changes arise. These two focuses of the controlling phase are categorized as performance evaluation and change control.

PERFORMANCE EVALUATION

This section covers performance evaluation topics. First, it offers a working definition of performance evaluation. Second, it discusses two general types of performance evaluation systems, with a focus on the more theoretical aspects of performance evaluation—what a project manager must have in mind when creating techniques to evaluate the performance of a construction project. Third, it covers specific methods of performance evaluation, from which project managers choose to implement in projects. Last, it discusses the risk management plan, which was developing during the planning phase, in light of the previous discussion on performance evaluation. The focus will be on how to use the risk management plan in selecting and developing performance evaluation techniques.

Primary Purpose of Performance Evaluation

The primary purpose of performance evaluation is to provide relevant and accurate data to project decision makers so they can make more informed decisions resulting in a higher probability of project success. Consider each component of this primary purpose:

Relevant and Accurate Data

Relevant and accurate data is a phrase loaded with both meaning and ambiguity. *Relevant* means that the information actually has a bearing on the situation at hand. Some information is interesting, but it may not be relevant. Other information may seem uninteresting, but it may be very relevant to the issue at hand. The project manager must be able to parse situations and information to see which parts are truly relevant to the decision-making process. *Accurate* means that the information being used is a reliable description of what actually transpired. If someone records incorrect data in the project plan, so that it does not accurately describe how the project is progressing, decisions based on this information might actually harm the project.

The project manager can help ensure that the information he is given is both relevant and accurate through a couple of means. First, he can explain portions of the decision-making process and the factors that are important to those members of the project team who are gathering the information. If people have a clear understanding of what they are attempting to accomplish, they are much more likely to provide relevant information. To ensure accuracy, he can do two things. First, he can make sure there is a system in place to access verifiable information. This means that as materials are ordered and as subcontractors perform their tasks, detailed records are kept and added to a portion of the project file. Second, he can make certain that project team members understand that they are not filing away pieces of paper for no particular reason. If people understand the importance of the information they are gathering, they are more likely to record accurate information.

Project Decision Makers

Project decision makers are the people with authority over various aspects of the project. The goal is to get the right information to the right people. This may mean that the information needs to get to the project manager, or it may need to reach certain key stakeholders. Whoever will be making a decision will need to have the information at hand. For instance, if a project manager has learned of an important trend related to project performance from his earned value (EV) evaluation, the results will need to get to whomever it is that can make a course-correcting decision. Information that does not get to the right person is useless. A good control system will not only state what information should be gathered, but it will also state to whom the information should go.

Informed Decision

Making an *informed decision* requires the two previous parts—relevant and accurate data in the hands of the appropriate decision maker. However, a third piece is required: using the information to make a decision. A decision that does not rely on the actual conditions of the project is likely to further deteriorate the project's chance of success. This is the moment in which the project manager's and the project team's experience and problem-solving skills show themselves. A project manager can implement proven project monitoring techniques, which will provide relevant and accurate data, and she can design a system that channels that information to the right people, but the information must still be analyzed, understood, and interpreted in such a way that good decisions are made.

Making good decisions is probably the most important part of the monitoring and controlling phase of the project lifecycle. The whole purpose of this phase of the project lifecycle is to gather information and transform that information into course-correcting decisions and process improvement. If a project manager does not possess the ability to translate the information into these types of decisions, his tenure as a project manager will most likely be short. It will not matter how well he can plan a project or how well he can execute various portions of it. It matters most how he can manage the hundreds of little parts that go wrong while the project is in progress. If he cannot continue to make course-correcting adjustments as the project is in progress, he will find that the project will not end as it should. It will either take too long, cost too much, or it will be deficient in quality in some manner.

Higher Probability of Project Success

In order to determine if the chance of project success is increasing, the project manager must know what the project is seeking to accomplish. The target of the project would appear to be to build a home or a garage or a commercial structure. However, that's not all there is to it. The project will not succeed if it does not build the intended structure, but it may actually build the structure and still fail miserably. The plan is not merely to build something, but the plan is to build a certain type of building for a certain cost in a certain time frame to a certain standard of quality. Success is measured not merely by whether the correct building was built, but whether it was done according to the plan that was developed and agreed on. If a builder were to build the right type of home, but at twice the estimated cost, this would hardly be called a success.

Informed decisions are not based merely on the accurate and relevant data that has been extracted, but also on where the project manager is directing the project. Decisions that are made in the present are based not only on the past, but on the desired future. They are made to adjust the direction so that the project is more accurately pointed toward the desired end point.

Having this purpose in mind will help the project manager when he is developing a plan for how the performance of the project will be evaluated. It will act as a guide when determining which measures and techniques will be used. The following factors should be considered when determining which techniques or measures to use:

1. How does one ensure that accurate and relevant data is collected?
2. Toward whom does the information collected need to be directed?
3. Does the decision rely on the information and does it contribute to project success?

Types of Performance Evaluation

There are basically two categories of performance evaluation systems: those that correct and those that improve. The distinction between these two types is not always clear, but they are helpful categories in as much as they focus on both the proactive and the reactive approaches to controlling the project.

Corrective performance evaluation techniques focus on how one corrects what has gone wrong. It is the reactive side of the equation. In general, corrective action is needed when the actual work of the project fails to conform to the planned work of the project in such a way as to threaten project success. The situations requiring some type of corrective action are numerous. The project may experience scope creep—the client continues to add work to the project so that it continually grows in scope and size. The project may begin experiencing cost overruns which threaten the success of the project. Or the project may encounter schedule problems for a variety of reasons, which could delay completion for weeks or months. The type of problem is not so much the issue; the issue is that the problem needs correction in order for the project to succeed.

In order for a corrective action to be necessary, a deviation must occur. In order to learn about a deviation, the project's performance must be monitored. Various methods of evaluating performance are discussed in the next section, but the important point here is that performance evaluation methods have the primary purpose of recognizing deviations from the project plan. The method that will lead to the discovery of deviations in an effective and efficient manner should be employed.

The first part is to recognize the problem or the deviation; the second part is much more difficult—developing a solution. Solutions come in all shapes and sizes. Usually the more simple a solution is, the more likely it is to work. The purpose of any and all solutions is to correct the root cause that led to the deviation.

The second general type of performance evaluation is process improvement. Process improvement is a broad term that can refer to many different levels of application in many different types of projects. It is applicable to construction practice as well as management practice. For the most part, the project manager will focus on management practices. He is not precluded from attempting to improve actual construction practices, if his knowledge and abilities allow him to do so; however, in this book, the focus in on improving the management of construction projects.

In the past 20 years or so, process improvement has become very fashionable, for good reason. It is called by any number of names: lean production, theory of constraints, benchmarking, business process reengineering, Six Sigma®, total quality management, and a number of others. All of these methods deal in some way with the topic of process improvement, and they all provide a lens through which one can study how work is being done. By looking through these lenses, a project manager should see ways that she can improve processes.

In process improvement, the goal is to do one's work in a more efficient and effective manner. The question is, how can work be done in less time, for less money, and be of a higher quality? Let's look at a particular method—the theory of constraints. The theory of constraints was developed by Dr. Eliyahu Goldratt and first introduced in his 1984 book *The Goal.* The book was primarily aimed at improving management practices in a production environment, but the theory was quickly applied to a wide range of disciplines. The basic premise of the method is that in every system there are constraints that hamper success. Through a process, the primary constraint is identified and dealt with in such a way that it is no longer the primary constraint. After this is done, the next primary constraint is identified, and the process is continually repeated.

This theory has direct application to project management techniques, and the project manager would benefit from studying it and other such methods. If the project manager is involved in a project in which he knows there is room for improvement, but he doubts that he has the time to master the various techniques and disciplines, a crash course in using the theory of constraints is offered in the next section.

Areas of Performance Evaluation

Before the methods of performance evaluation are discussed, let's think through the various areas of the project on which a project manager focuses during performance evaluation. When developing the project plan, the project manager mapped out the entire project from start to finish. She set the budget and the schedule, and she described the extent and the quality of the workmanship. These are the three main areas of the project that are monitored during the controlling phase of the project lifecycle. As well, the project manager is managing the interpersonal aspects of the project to make certain that any issues among those who are working on the project do not lead to problems in one of the other three areas. The primary areas that the project manager is seeking to control throughout the execution phase of the project are:

1. Cost
2. Schedule
3. Quality
4. Interpersonal

The focus at this point is not on the various tools that can be employed to control the different aspects of the project, but rather on developing a perspective for how one should approach monitoring the project. In the upcoming section of the chapter, specific tools will be described and their application will be presented.

Cost Control

During the planning phase of the project, the project manager completed a cost estimate for the entire project, which was then used to develop a project budget. The initial project budget becomes the cost baseline for the entire project. The cost baseline acts as a standard to compare the actual work of the project as work progresses. The cost baseline is only rarely changed. It can be changed when a major change is implemented in the project. But it does not change merely because something was more expensive than anticipated. For instance, if the client decides to add a bonus room above the garage, the project manager prepares a change order, which would show the impact of the change order on the cost, the schedule, and the quality of the work of the project. Once the client accepts these changes, the baselines are updated to manage such a change. However, if the client decides to change the color or style of her tile, the manager most likely does not change the cost baseline.

In order to effectively control the costs of the project, the project manager must focus on a couple of things. First, the project manager must ensure that the project plan is followed. The vendors who won the bids based on a certain estimate must be used. If a project manager quoted a framing package during the planning phase for a certain price, but then decided to use a different vendor at a higher price, he is possibly making a foolish decision that might lead to cost overruns. The cost overrun would be due to the fact that the project manager failed to follow his own plan. This can also happen when a subcontractor attempts to charge a higher amount because she underbid the project to get the contract. In this situation, the project manager must not give in. He must hold firm to the agreed on price. If he fails to do this, he can expect cost overruns.

The project manager should not only focus on following the plan, but also on controlling the quality and the schedule of the work. These aspects will be discussed more fully in the next section, but they are related to cost control. If the quality of the work declines, rework will be necessary or the project manager may encounter litigation. If the project manager falls behind schedule, then he may have to pay a penalty for a late project. The point is that every area of the project affects every other area of the project. It is a ripple effect that can have devastating consequences.

There are specific tools to control the cost of the project. The project manager should have a solid purchasing system that ensures that costs are within budget. She can also use earned value management techniques to project cost overruns and she should focus on making certain that the project is on schedule and that the work is done properly. These and other specific methods will be discussed in the next section of the chapter.

Schedule Control

Controlling the project's schedule is one of the most difficult aspects of the construction project. In many manufacturing settings, the work is being done in a controlled environment using automated processes. The construction manager does not have the benefit of a controlled environment. She must deal with the weather, material shortages, vendors, subcontractors, and a host of other problems that can wreak havoc on the project's schedule. This has led some project managers to not even develop a formal schedule. They know what work needs to be done, and they just move from one task to the next with only a general idea of when they will finish. Typically, the more customized the work, the more difficult the schedule is to maintain.

One of the purposes of this book is to emphasize the numerous benefits that planning on the front side of the project can have. A major part of the planning phase centers on developing a realistic construction schedule that is actually workable. This is not to say that the schedule developed during the planning phase will be followed perfectly; it will need to be changed. But each time that a project manager develops a construction schedule, she will be better qualified to create a more workable and realistic schedule that includes the variety of issues that arise during the construction project.

Controlling the schedule, at a minimum, includes two things. First, controlling the schedule means working to follow the schedule that was developed during the planning phase. The work was ordered in a specific way for specific reasons, and the schedule should, given unforeseen circumstances, be followed. The estimated time required to perform the work was also based on reasonable premises, so it is reasonable to assume that the work will be done in the time allotted. Sometimes keeping to the schedule means that the project manager will need to prod, push, or even pull the work along, but that is part of the project manager's responsibility.

Second, controlling the schedule means updating and revising the schedule as the work progresses. It is going to rain; rework will be needed. As work progresses, the project manager adjusts the schedule to the realities of the project. If the project manager fails to do this, she will only be creating more confusion, which could further delay the project.

There are a number of techniques for managing the construction schedule. One that this book will propose later is earned value management (EVM). EVM allows the project manager to measure progress based on the schedule baseline and predicts future performance. Other techniques and tools will also be discussed later in the chapter.

Quality Control

The quality of the work is also a major area of the project that must be controlled. Two primary aspects of quality control are involved. First, quality control refers to the quality of the workmanship of the home. Is the home being built according to an acceptable level of skill and finesse? Is the work being done according to the specifications outlined in the contract? Second, quality control can refer to the skill with which the project is being managed. Is the project being managed well? Is the project manager using project management practices that are most likely to lead to success? The project manager is concerned with both aspects of quality control. She must make certain that the actual work of the construction project is performed according to the standards set forth in the contract and the building specifications. She must also make certain that the way the project is managed does not lead to mistakes, schedule delays, or cost overruns.

The project manager controls the quality of the workmanship through inspections. She must inspect the work as it is done to ensure that it meets specifications outlined in the project plan. In the chapter dealing with project execution, the requirement to inspect and approve work was repeated in almost every detailed section of the chapter. This may have seemed repetitive because it was repetitive, but it was repetitive by design. The objective of the chapter is to reinforce the idea that one should work the plan and inspect the work—the key to success in construction management. One of the best tools that a project manager can use to inspect work is a checklist. Although it may seem a little old-fashioned, there are few things much better than creating a list of things to check at each stage of the construction. Checklists must be customized for each construction project, as each project has its own design and challenges. But the purpose of the checklist is simply to make certain that the work is performed in conformity with the specifications found in the project plan. Checklists can be created fairly easily and offer a great tool for ensuring quality workmanship.

The project manager is also concerned with making sure the project management meets a high standard. Regardless of the quality of the plan, if the project is poorly managed and executed, the project is destined for delays, cost overruns, and various amounts of rework. The project manager can avoid this by ensuring that the procedures for ordering materials, scheduling subcontractors, issuing payment, and all the other aspects of the project are followed as outlined in the project plan. Poor management typically shows itself in poor workmanship; a project that is poorly managed will be poorly executed.

Interpersonal Control

When writing this section, I struggled between calling it *Interpersonal Control* or *Conflict Control*. I picked interpersonal control because it sounded less negative, but conflict control may actually be the more correct title. There are generally two types of conflict: constructive and destructive. At one point in my life, I worked as a business analyst on a software project. Over the year or so that I was engaged in that project, I never endured more conflict in the work environment, and yet it was one of the most enjoyable jobs I ever did. The reason it was so enjoyable was that the conflict was constructive. It made me think, defend my ideas, and develop a discerning ear. When someone said that he thought the other person was wrong, it was not an attack on the person, but on the validity of the idea. The other person would have to defend her idea, rebut concerns, and explain potential problems. It led to a very solid piece of software.

Some people might find constructive conflict to be a totally foreign concept. For them, conflict is negative, and it might always be a negative experience. Many times project managers do not invite critical thinking, but only want those who approve without question around them. This is incredibly destructive to the life of any project.

The project manager should understand the difference between constructive and destructive conflict and should seek ways to foster constructive criticism and deflate destructive conflict. Depending on the people involved and the environment, this will seem nearly impossible in many settings. Each project team has its own set of dynamics. But regardless of the specific dynamics, all teams share some things in common—they go through similar phases of team development: forming, storming, norming. Later in the chapter, specific techniques for dealing with the various dynamics of teams and conflict will be discussed.

Methods of Performance Evaluation

Now that the various areas that need control have been discussed, let's focus on the specific tools and techniques of performance evaluation and their applications. First, methods of corrective action are discussed. This focuses on how one goes about correcting problems that have arisen. Second, methods of continuous process improvement are briefly discussed with particular attention given to the theory of constraints.

Methods of Corrective Action

The project manager has invested a great amount planning, research, and effort in developing the construction project plan. She has spent many hours attempting to discover the bumps that might lie ahead. The plan may have been reviewed numerous times and scrutinized from every angle to find potential problems. But given all that planning, all that hard work, every project manager worth her salt knows that problems are going to arise and that those problems are going to have to be solved. This undeniable truth is why that oft-repeated paraphrased line from Robert Burns' poem "To a Mouse" (1786) rings so true in the ear of every project manager, "the best laid plans of mice and men often go astray." When things go wrong, corrective action is what is needed. The project manager must meet these issues head on with solutions. The project manager who can develop solutions is the project manager who stands a good chance of succeeding in the long run.

In this chapter, two categories of strategies or tools for correction will be discussed. The first are the more general strategies and tools, which have applications across a broad range of uses. They will provide insight and help when dealing with scheduling issues, budgeting issues, quality control issues, interpersonal issues, and most other types of issues that might arise during the construction project. The three general tools are:

1. Root cause analysis
2. Earned value management
3. Risk management plan

After these three general tools are discussed, specific strategies for each of the four primary areas of concern will be discussed.

Root Cause Analysis

In order to solve a problem, the project manager must first understand the problem. In graduate school, I took a class on business strategy. The class focused on studying various companies which had suffered devastating failures at certain points in their history. Sometimes it was the failure of a specific initiative or project (Crystal Pepsi), and sometimes it was the failure of the entire company (Washington Mutual). The students were given a case file for each project, which contained a large number of details. Some details were insignificant, whereas others contained very significant clues to the failure. The point was to determine from all the details what was significant and what was insignificant. There were many symptoms, but what was the root cause?

The hugely popular television drama *House* centers around a brilliant but complicated doctor named Gregory House and his team of doctors who specialize in solving the most difficult and complex medical cases. In each show they must consider all the symptoms and determine the root problem. What medical issue is causing all of these symptoms? If they are unable to determine what root cause is manifesting itself through the various symptoms, the patient will die; but if they are able to find it in time, they will be able to save the patient's life.

In the business strategy course, a student's failure to find the root cause would lead to a poor grade. In the television show, the situations are scripted, and the doctors always find the cure just in time. However, for the project manager reading this book, failing to find the root cause could result in the project failing and the manager suffering a range of disciplinary actions. The project manager's ability to determine the root cause of a problem through effective problem solving is critical for success.

Methods for searching out root causes are as varied as the people using them. If one were to interview 100 project managers and ask them how they go about discovering root causes, a number of methods would emerge. Most of the time, the answer likely would not reflect a formal system, such as a "Seven Steps for Successful Root Problem Identification." Instead a mostly organic approach would be described, in which the project manager sees an issue arise and then begins an informal process of uncovering the cause. Most project managers who see the problem and attempt to understand the cause of the problem will typically locate the root cause eventually. But the longer the problem goes uncorrected, the more devastating the effects can be on the project. Having an actual method for investigating root causes can be an advantage. The method might need to be modified, depending on the application, but the basic process should serve most projects.

The steps moving through the root cause analysis are:

1. Identify symptoms
2. Determine rate of recurrence
3. Determine root cause of symptoms
4. Succinctly state root cause
5. Develop alternative solutions
6. Implement solution(s)

Identify symptoms Identifying symptoms should be a relatively straightforward process, for it is the symptoms that lead the project manager to the awareness of a problem. It would be nice if it could be stated that project managers are always aware of the symptoms that may indicate deeper problems, but this is not always the case. Many times the issue is raised by a project team member or a stakeholder, such as the client. However it comes to the attention of the project manager, it is best to write out the problem to ensure that it is clearly understood.

This may seem to be a rather rudimentary recommendation, but writing out the problem does a couple of things. First, it makes certain that the project manager and others involved in the situation have a shared understanding of what the problem actually is. By crafting this statement together, the group greatly reduces the possibility of continually talking past one.

The second benefit of writing out the problem is that it separates true problems from what one might call differing management styles. What a new team member believes to be a major problem might not be a problem at all for the project manager. For instance, the project manager might prefer to install the unfinished hardwood floors throughout the home before the cabinets are set and the fireplace is built. But a new team member might have previously worked at a company which installed the hardwoods much later in the project to avoid damaging them. This might seem like a major problem to the new team member, but in fact, it is not. This is a fairly simple example, but it points to the fact that what is a problem for one person is considered a best practice for another.

In identifying the symptoms, those involved might identify only one or they might identify a few. The ones identified might seem to be directly related or completely unrelated. Many times the connection between various symptoms is not immediately visible, but through careful consideration and investigation, the connections, if any, will become apparent.

Determine rate of recurrence Once the symptoms have been identified and agreed on, the next step is to see if these are new symptoms, or if they have been present throughout the entire project. This is an important step, as it offers insight into the source of the problem. If, for instance, over the past few weeks, material orders have begun to arrive late, the project manager might look to recent weather conditions or untypical problems at the supplier source. If, in fact, the materials have been arriving late throughout the project, but it went unnoticed, the project manager must determine two things: why the materials were late, and why no one noticed. The rate of recurrence will give insight into whether the problem is related to a faulty management system, or if it is due to simply failing to follow a proven method or system.

After the immediate problem has been investigated, the project manager might find it helpful to go back and consider previous projects. Was this a problem that was experienced on multiple projects? If so, what was done to correct the problem in previous projects? This is why it is so important for the project manager to complete lessons learned reports throughout the life of the project. These types of reports provide a unique snapshot of the project at various points so that future project managers will have a clear record of the project, instead of relying on the selective memory that many suffer from.

Determine root cause of symptom(s) Determining the root cause of the symptom is easier said than done. In general, there are three categories of root causes:

1. Acts of God
2. System failure
3. Human error

Acts of God

The first category is the one that project managers can only attempt to manage through prayer: rain, snow, tornadoes, hurricanes, earthquakes, flooding, and any other natural event can create all sorts of problems for projects. Schedule delays, materials shortages, labor shortages, and many other problems can result from the weather. If the problem is a delayed construction schedule, and it has been raining so often that the foundation cannot be built, then short of erecting a tent over the site and finding a way to dry the ground, there is little that can be done. This is the easiest root cause to identify, and if it really is the root cause, the project manager is only held responsible by the most unreasonable of clients.

System Failure

A system failure occurs when the method of managing the project is faulty. The project team members faithfully follow the protocol and guidelines of the project plan, but the symptoms are still present. In this scenario, the problem is that the method does not work. It may have looked as though it would during the planning phase, but, in reality, it does not. This is often difficult for a project manager to recognize, because he is typically the person who has developed the system. He is more likely to assume that the project team is simply failing to follow the plan in some way. An example of this might be that the company for which the project manager works has a particular protocol for approving material orders. This protocol may not have been adequately accommodated when scheduling the lead time for materials, which might lead to delayed shipments. Or it could be that the company has a specific time frame for issuing payment for work completed, which may be longer than subcontractors want to wait, leading to subcontractors lodging complaints to the project manager. In the above scenarios, the project manager may not be able to change the company's policy, but he can begin ordering his materials earlier, and he can better explain the payment waiting period to subcontractors before they agree to do the work. Either approach might possibly solve the problem. The point is that the system is faulty; correction requires either changing the system or altering the plan to be better integrated with the system so that any delays are taken into consideration.

Human Error

The third general category of root causes is human error. Human error is a broad category with any number of subcategories. If a project manager is working on his first few projects and has not had much guidance, there is a good chance that problems are attributable to a poor system. If, on the other hand, the project manager has developed a proven system of managing construction projects, deviations from the plan are more likely attributable to human error of some kind or another. In general, human error is attributable to either a lack of understanding of the process or plan or a lack of skill necessary to perform the work.

When a problem is determined to be due to human failure, the first assumption is that the person(s) responsible for the problem simply failed to understand some detail about the management plan. If this is the case, many times the problem can be easily remedied by informing the individual as to the proper way of doing the work. This might be the case when the problem is that material vendors are receiving late payments for material orders. The person responsible for approving payment might be incorrectly requesting payment from the accounts payable department, which is creating the delay. In this instance, the person must be informed of the proper procedure; this typically corrects the problem. However, this scenario assumes that the person possesses the functional and behavioral skills necessary to perform the work.

If a worker lacks the functional skills, without training the person is unlikely to be able to perform the work even if the correct procedures are explained. This concerns aptitude and ability. Does the person who has been placed in a certain position possess the aptitude and technical ability to perform the work that has been assigned? If not, the individual has been misassigned within the project and must be reassigned.

This typically occurs when a worker has been promoted to a new area of responsibility that requires a higher level of skill. The candidate appeared to possess the necessary skills based on previous experience with lesser roles, but given the new role, he is unable to rise to the challenge. This may occur when an assistant project manager is given his first project as the head project manager. Some people make excellent assistant project managers, but lack the ability to lead the project on their own. The project manager has the responsibility to determine if her project team members are capable of performing their jobs at an acceptable level. If she determines that they lack the necessary skills to perform the job without error, she must reassign them.

The second type of skill that is necessary to perform the work is behavioral or interpersonal skills. Interpersonal problems are among the most undertreated problems on the construction site. One subcontractor becomes angry at another over some minor issue or insult, and the project suffers. A project coordinator is too abrasive or rude with subcontractors, which leads to subcontractors arriving late or cancelling the contract. The project manager must be attuned to the fact that these types of problems can arise among the various stakeholders, and he should be a solution, not a cause, of these types of problems. This requires attempting to see the forest without focusing too much on the trees. The project manager should take a pragmatic approach to managing the issues with the view that completing the project according to the project plan is the goal, not playing Dr. Phil or Jerry Springer. He needs to manage the various competing personalities with the goal of making sure that everyone can set aside personality conflicts and get their work done on time, within budget, and according to the quality standard set forth. This is not to say that the project manager should run over everyone, but she should navigate through the various interpersonal conflicts in a way that is appropriate to the situation. Sometimes this will require giving ground; other times it will require taking ground. The wise project manager will recognize the time for each.

After considering various symptoms in light of the previous discussion, the project manager should be able to succinctly state the root cause of the problem, which is the next step in the process.

Succinctly state root cause Once the symptoms have been identified, the rate of recurrence has been determined, and the various sources of root causes have been investigated, the project manager should be equipped to succinctly state the root cause of the problem. Actually stating the root cause is important for the same reasons as stating the symptoms: to ensure mutual understanding and agreement. By making a statement tying the previously identified symptoms to the root causes, the project manager obliges those working to correct the problem to reach some type of consensus through verbalized agreement.

Making this statement and achieving agreement is essential before moving to the next step for a few specific reasons. First, if this is not done, it is less likely that everyone shares a common understanding. By writing out the symptoms and root cause, it is more likely that everyone is defining the terms the same way. Second, it helps ensure that everyone is working on solutions to the same problem. Even though the process being described progresses through linear stages, the mind typically works differently. It seems to me that one is not typically able to think through the various types of root causes without selecting one and immediately moving through the various options to correct the problem. Even in writing this section of the book, I have offered possible solutions or root causes before that stage of the process has been reached. It is natural for the mind to do this. Because of this, someone might connect with a certain idea early on and see that as the obvious solution, which may cause her not to carefully consider the information that is later developed. If this is the case, then she might arrive at a conclusion much different from conclusions reached by those who had a more open mind during the process. Open discussion leading to the drafting of a statement on the problem serves the purpose of helping to ensure agreement on the problem.

Develop alternative solutions One might wonder why the title for this subsection is not *develop a solution*. Typically a variety of solutions are possible for any given problem. The solution one chooses will be determined by any number of factors.

Consider the following example. On a construction project in which I participated, we discovered at the end of the project that the initial lot survey performed to locate the foundation of the home was incorrect, which caused the home to be built within the building setback. Instead of being 12 feet off the side property line, the home was only 6 feet off, creating a structural encroachment within the setback. This was discovered when the client had a new survey completed in order to determine where fencing and landscaping could be placed without encroaching on the neighboring lot. The buyer was purchasing the home for cash, so there was no approval necessary by an outside body before the buyer could close on the property. This problem had a few possible solutions.

A first solution, which was not desirable for either party, was that the clients accept the job *as is* with a comment on the title insurance policy noting the encroachment. This would be the most costly solution to the client, because if she decided to sell the property in the future, she would have to disclose this fact. A second option was to get a variance from the county, which would in effect make the encroachment legal. This is a better solution than the first one for the client, but still is not the best. Next, a portion of the neighboring lot could be purchased and added to the client's lot, which would effectively move the setback line so that the home was no longer in violation of the zoning code. This is one of the better solutions because it removes the problem entirely. However, the neighboring lot owners had to be willing to sell a portion of their lot. A last option considered was to purchase the entire lot next door, then reapportion a section of the neighboring lot to make the client's lot comply with the zoning code. For each of these various options, there are different considerations and barriers that must be addressed and overcome. The project manager had to work with the various stakeholders to develop a solution or risk a lawsuit or other negative consequences.

As one can see from the previous example, a variety of solutions might exist depending on the type and complexity of the problem. Many times the project manager prefers one solution over another for any number of reasons, but she is unable to implement it because of the competing needs and desires of the various stakeholders. The project manager's company is typically looking for the least costly solution that will comply with the terms of the contract and satisfy the client. The client is looking for the solution that will be most advantageous to her. Other stakeholders have other competing needs that might conflict with those of both the project manager and the client. The more complex and far-reaching the effects of the problem, the more stakeholders the project manager will be required to manage. Developing a solution that satisfies everyone will not always be possible, so the project manager must be ready to make tough decisions and handle the backlash.

Implement solution(s) Once a solution has been chosen, the project manager will have to implement it. The magnitude of the solution to be implemented will determine the amount of time necessary to plan such a solution and the amount of monitoring necessary to ensure that the solution is properly working. For instance, if the needed correction is merely placing material orders earlier, it should be fairly simple to implement the change. However, if the project is behind schedule, and some intense scheduling is necessary to speed up construction, the project manager must implement the solution more quickly and more closely monitor the changes.

It is imperative that the project manager monitors the implementation of the solution to ensure that the corrective action resolved the root cause. This is evidenced by symptoms disappearing and not recurring. If, after the solution has been implemented, the symptoms are still present, either the solution has not yet had its desired effect or the selected solution will not actually correct the problem. In this case, the project manager must make the necessary changes to tweak the solution or attempt another.

A key component of this entire process is documenting the process and the findings. If a project manager takes the time to document detailed lessons learned, then once a project manager has done a few different types of projects he will have a wealth of historical knowledge to draw from on future projects. The responses might need adjustment for the particular needs of a new project, but at least he will have a starting point.

The investigative process discussed previously is applicable to any of the major areas of the project that requires control (cost, schedule, quality, interpersonal). The basic process will remain the same, but the tools used in the process and the solutions applied will vary, depending on the type of problem that one is addressing. For instance, the quality control checklists that the project manager uses to manage the quality of the workmanship are unlikely to tell her how far behind schedule she is or what the project cost of completing the project is. It might provide insight into why there are schedule delays or cost overruns, but this will be secondary information. If the project is behind schedule, specific tools are available to learn how far behind the schedule is and different solutions can be employed to attempt to get the project back on schedule. Therefore, the remainder of this section will provide a brief overview of a few of the methods one can use to both identify and correct problems. The first method to be discussed is called earned value management.

Earned Value Management

Earned value management (EVM) is a project performance measurement and monitoring tool, as well as a forecasting tool. More formally stated it is "a management methodology for integrating scope, schedule, and resources, and for objectively measuring project performance and progress" (The Project Management Institute 2008, 209). Another way of stating this is that EVM objectively depicts the relationship between scope, schedule, and resources in a particular project so as to show how well the project is performing and is expected to perform in the future. It shows these relationships by the use of some rather simple mathematical ratios or formulas. The project performance is measured or depicted by determining the budgeted cost of work performed (i.e., earned value) and comparing it to the actual cost of work performed (i.e., actual cost). Progress is measured by comparing the earned value to the planned value (The Project Management Institute

2008, 209). Although the calculations are rather simple, as is the concept of EVM, it is a challenge to fully leverage EVM.

EVM is a scalable tool that can be used on a project of almost any size. It is assumed that most residential construction projects are relatively short (less than one year) projects. These projects will gain much benefit from merely using the standard EV metrics, which will be discussed shortly. However, the term *earned value management system* (EVMS) refers to a much more formal and comprehensive system that "at minimum complies with the 32 criteria as set forth in the American National Standards Institute Guidelines, ANSI/EIA-748-1889. Compliance with the EVMS criteria has historically been required by the Department of Defense, such as in DoD Instruction 5000.2 and has been mandated in many cases by agencies such as the Department of Energy (DOE), the National Aeronautics and Space Administration (NASA), and the Office of Management and Budget (OMB)" (Budd 2005, 38).

One of the primary reasons residential construction project managers should consider EVM is that it is a natural fit to the residential construction industry, for either custom builders or production builders. Almost all builders, regardless of the formality of their project management, use some form of a work breakdown structure (WBS), although they may never call it such. Whenever a builder either bids on a job or prepares to build a spec home, the house's specifications must be outlined and costs estimated. Along with the cost, an amount of time is estimated for each task in the building process. Custom builders will do much of this from scratch for each project, whereas production builders may have a template prepared, which only needs to be updated. This specification sheet or job card or construction checklist acts as the WBS, as it guides the construction process; it also creates a cost and schedule baseline for the project.

As most builders use this type of tool and are interested in comparing actual performance with planned performance, EVM is a natural fit. The information necessary to perform EV analysis is already being collected by most residential contractors. Most residential construction companies could begin benefiting from EVM more easily than many other industries might. The primary obstacle to overcome is education. Most builders either lack knowledge of the tool or they lack the expertise to begin benefiting from such a tool.

Not only is EVM a natural fit for residential construction companies, it also provides a number of benefits. First, it creates a standard way to compare project performance regardless of project size, and it provides a common performance appraisal language (Barlow and Klingelhoets 1997, 61). Many times builders need a way to determine which project to focus the most attention on, or which project is falling behind schedule or is going over budget. By using the proven quantitative tools of EVM, project managers can compare projects more easily. Projects that are beginning to slip in one area or another can be isolated, and the necessary attention can be given to those projects. This can be an automated process for those who use project management software. Many standard software packages, such as Microsoft Project, have these tools built in; users merely need to be made aware of how to use them and how to interpret the data correctly. If a standard software package is not available, one could easily create an EVM template in any spreadsheet application, such as Microsoft Excel.

Not only does EVM create a universal way to view project performance, but it will also allow the builder to estimate the future project performance based on past performance. In my experience, those within the construction industry, like those within other industries, tend to be primarily reactive, rather than proactive. They assume that they have no way of forecasting what the future will hold for their project, but EVM provides a solution to this problem. Granted EVM will not tell them if the price of lumber or concrete will increase, but it will tell them with relative certainty if their project is going to go over budget. This could be very advantageous to builders who are working under both fixed-price and cost-plus contracts. For those working under fixed-price contracts, it tells them that their profit margin could be decreasing by the day. For those with cost-plus contracts, it alerts them to the fact that the future homeowner will be unpleasantly surprised. Most importantly, it alerts the builder that some corrective action, if possible, must take place, which brings us to one of the weaknesses of EVM.

Once the project manager has forecasted either a schedule overrun or a negative cost variance or both, some action must be taken to determine how the project can be brought back into control. EVM does not tell the builder what action needs to be taken; it merely alerts the builder to the fact that an undesirable variance has occurred, which if unchecked could spell ruin for the project. It is important for the builder to realize that some variance is to be expected—maybe some small percentage of float between the negative and the positive—but if the trend is continually moving toward the negative side, some action must be taken. EVM only reports the relationship between what actually happened and what was planned, not how to reconcile the difference.

Yet even here, EVM has benefits that can greatly help the builder. EVM is a scalable tool, meaning that it can analyze the project on varying levels of detail (Barlow and Klingelhoets 1997, 61). One may look at the project as a whole or at a specific area of the project. When a negative variance exists, the project manager can begin moving down into the detail of the project to determine what caused the variance to occur. Once again, EVM does not say how to correct the variance, but it can isolate the source of the variance, thus improving the project manager's probability of correcting the variance quickly. As with any performance measurement tool, the human element is critical, which is why proper training is essential to EVM success. Sometimes companies must seek the assistance of an experienced practitioner to help them navigate the sometimes rough waters of EVM.

The fact that EVM is a scalable tool leads to another benefit that many builders may find very helpful. By using EVM to manage the construction project's performance, builders will have a new method by which to analyze subcontractors' performance. EVM can be applied to the work performance of subcontractors, thus allowing project managers to know for certain which, if any, subcontractors are causing variances in the project. It is important to note, however, that qualitative factors [honesty of subcontractor, quality of work] must be considered alongside of the quantitative performance measures provided by EVM. Although EVM cannot measure these qualitative factors, it does provide a means of pinpointing subcontractors who fail to follow the contract schedule or the terms of their contracts. This is especially beneficial for managers seeking to establish long-term relationships with subcontractors. EVM analysis can inform builders which subcontractors are the most reliable in their performance, thus giving the builder more certainty when bidding projects.

Another, possibly surprising, benefit of EVM is that it can expand the effectiveness of one's risk management plan, as these are complementary tools. Although EVM is a reliable tool to forecast certain areas of project performance, its true strength lies in its ability to analyze past project performance (Hillson 2004, 1). On the other hand, risk management is primarily concerned with projecting both the probability and impact of future events occurring during the project's life. One primarily looks back, whereas the other primarily looks forward; yet both are focused on keeping projects performing according to the project plan. Once EVM identifies a variance and isolates the cause of the variance, the risk management plan addresses how the issue will be handled. In addition, the forecasting portion of EVM can be used throughout the project to continually reference the risk management plan, determining if any indicators exist to initiate a portion of the risk response plan. Thus these two tools work together, each benefiting from the strengths of the other.

EVM is not without its pitfalls or its detractors. One weakness mentioned earlier is that EVM does not tell how variances should be corrected, which is one reason there is a natural link between EVM and risk management. EVM is also a poor motivational tool, which can lead to unreliable EV indices. If project managers are being rated primarily on the EVM ratios, this could lead those project managers to develop means of tweaking the indices in a positive way, resulting in an inaccurate representation of actual project performance. It is important that steps be taken to mitigate this risk. Such steps would include using EVM as only one of several indices to measure performance.

EVM results will only be as good as the baseline. If a poor project baseline is prepared, the results of EVM will be poor, which will lessen its usefulness as a performance measurement (Barlowand Klingelhoets 1997, 65).

Last, EVM is based on simple mathematical truths, but it is not necessarily a simple tool to implement. As with any new tool, there is a certain learning curve involved. Many companies have taken years to reap the full potential of EVM; the would-be user must realize that EVM is not without its challenges. Fleming and Koppelman (1999) have a series of articles summarizing the steps to implementing an EVM system, which is referenced at the end of the chapter; these documents are highly recommended as an overview of the EVM implementation process, but much more research and education will be necessary to fully utilize its capabilities.

A practical, real-world example might be helpful at this point. The example project is an actual construction project performed by the construction company for which I work. The project WBS has been simplified, as a detailed WBS would require more space than is available. Table 5.1 presents a high-level WBS of the project, with appropriate baselines.

As shown in Table 5.1, the project was slated to last 30 weeks, and the total construction cost estimated at $113,000. Although this is longer than should be required to build such a small home, the company had more than 30 other projects going at the same time, which extended the time it needed to complete this project. Let's assume that the first EV analysis (EVA) was performed at the completion

Table 5.1 Project baselines

WBS item	Baseline duration	Baseline cost
Lot prep and foundation	3 weeks	$20,000
Framing and masonry	5 weeks	$25,000
Rough-ins	3 weeks	$15,000
Septic and water	2 weeks	$10,000
Drywall, paint, and cabinets	6 weeks	$15,000
Interior finishes	4 weeks	$12,000
Floor coverings	3 weeks	$6,000
Decks, guttering, and driveway	2 weeks	$5,000
Grading and landscaping	2 weeks	$5,000
	30 weeks	$113,000

of week 8, when the framing and masonry should have been completed. Table 5.2 shows the project's status at the end of week 8.

As can be seen in Table 5.2, the planned value (PV) at the end of week 8 was $45,000 ($20,000 + $25,000), which was the budgeted cost for the work to be completed at that point in the project (The Project Management Institute 2008, 160). The actual cost (AC) incurred at that point in the project was $47,000. Earned value (EV), which is "the budgeted amount for the work actually completed during a given time period," is $45,000 (The Project Management Institute 2008, 160). EV is probably the most challenging number to determine, as the project manager must determine the value of the work actually accomplished. For instance, at the end of week 3, the grading was not yet complete. The grading contracting required an additional three days, at a cost of $500 per day. However, at the end of the third week, $18,000 had been disbursed, with only the extra grading remaining. To determine the EV at this point, one should take the PV of $20,000 and subtract any additional monies that need to be spent to accomplish the planned work. In this case, an additional $1500 was needed to finish the grading; the EV was calculated by subtracting $1500 from $20,000, which made the EV $18,500. By the end of the eighth week, the grading had been completed, making the total earned value $45,000.

Using the information from Table 5.3, the project manager can use EV metrics to analyze the project's performance to date, as well as forecast future performance. First, the standard EV metrics are used to analyze the current progress made in the project. Table 5.4 shows the relevant formulas to be used during this phase of the analysis.

The cost variance (CV) for the construction project is calculated by subtracting the AC of $47,000 from the EV, which is $45,000, giving a difference of −$2000. This shows that the project is currently $2000 over budget. The second calculation to perform is the schedule variance (SV), which is calculated by subtracting PV from EV, which in this case is $0. It can be inferred that the project is currently on schedule. However, if this calculation had been done at the end of week 3, the calculations would have shown us that the project was behind schedule. The last two items to calculate are the Cost Performance Index (CPI) and the Schedule Performance Index (SPI):

$$CPI = EV / AC = \$45,000 / \$47,000 = .96$$
$$SPI = EV / PV = \$45,000 / \$45,000 = 1$$

If either of these numbers is less than 1, it means there is a negative variance. If they are greater than 1, there is a positive variance. In the case of this particular project, it is known that the project is on time but currently over budget. Therefore, the

Table 5.2 Baseline variances 01

WBS item	Baseline duration	Actual duration	Baseline cost	Actual cost
Lot prep and foundation	3 weeks	3.6 weeks	$20,000	$19,500
Framing and masonry	5 weeks	5 weeks	$25,000	$27,500
Rough-ins	3 weeks		$15,000	
Septic and water	2 weeks		$10,000	
Drywall, paint, and cabinets	6 weeks		$15,000	
Interior finishes	4 weeks		$12,000	
Floor coverings	3 weeks		$6,000	
Decks, guttering, and driveway	2 weeks		$5,000	
Grading and landscaping	2 weeks		$5,000	
	30 weeks	8.6 weeks	$113,000	$47,000

Table 5.3 EVM calculations 01

Status	End week 8
Planned value (PV)	$45,000
Earned value (EV)	$45,000
Actual cost (AC)	$47,000

Table 5.4 EVM formulae

Cost variance (CV) = EV - AC
Schedule variance (SV) = EV - PV
Cost performance index (CPI) = EV / AC
Schedule performance index (SPI) = EV / PV

project manager should begin to inspect what caused the variance and what can be done to ensure that future variances do not occur.

EVA may also be used to forecast future project performance. Because the project is currently experiencing a negative cost variance, the project manager should forecast how this will affect the project if this trend continues. The formulas used to forecast future project performance are outlined in Table 5.4. The estimate to completion (ETC) is the estimated cost required to complete the project. For this particular project, based on the current progress, it is calculated:

$$ETC = (BAC(\text{budgeted at completion}) - EV) / CPI$$
$$= (\$113,000 - \$45,000) / .96 = \$70,833$$

Using ETC, one can estimate what the total project cost will be, based on current project trends. This is called the estimate at completion (EAC), which is an estimate of the total cost at the end of the project. It is calculated:

$$EAC = AC + ETC = \$47,000 + \$70,833 = \$117,833$$

If the project continues along the same trend as it has thus far progressed, the project will be $4,833 over budget. As mentioned in Table 5.5, the alternative formulas provide a more pessimistic account, as they consider both SPI and CPI in their calculations. However, in this case, there would be no difference, as there is no schedule variance currently in the project. If there were a schedule variance along with the cost variance, then the forecasted EAC would be higher.

Table 5.5 EVM forecasting formulae

Estimate to completion (ETC) = (Budget at completion (BAC) - EV) / CPI
Estimate at completion (EAC) = AC + ETC
Alternate ETC = (BAC - EV) / (CPI * SPI)
Alternate EAC = AC + Alt. ETC

Now move to week 21 of the project. The pertinent project information is shown in Table 5.6. As can be seen in Table 5.6, the project is progressing with some tasks being completed on time and some being completed both behind and ahead of schedule; yet from such a chart, it is difficult to see how the variances will affect the project as a whole, which is why EVM is so helpful. EVM allows the project manager to analyze the project data to date and see what the effect will be.

Once again, the analyst must identify the key values to perform EV analysis, which are EV, PV, and AC. EV and PV are not easily discernable from the chart. The analyst must speak with those performing incomplete tasks to determine how much value has currently been earned by the work completed and how much value was planned. On consulting the interior finishing crew, the analyst determined that the EV to date in their task was $4000, which corresponds to actual cost to date. She also learned that the PV was $8000 to this point in the task. With these numbers, she can add the EV for interior finishes of $4000 to all the previous EV, which corresponds to PV ($20,000 + $25,000 + $15,000 + $10,000 + $15,000 + $4000 = $89,000), for an EV of $89,000. She may also add the $8000 PV to the previous PV for all previous weeks ($20,000 + $25,000 + $15,000 + $10,000 + $15,000 + $8000 = $93,000). These key numbers are provided in Table 5.7. From these key numbers, she can perform EV analysis to gain a better understanding of the project's current performance. The results are provided in Table 5.8.

Based on this last analysis, the project manager would be able to see that the project is now both behind schedule and over budget. The forecast provided by EV shows that the project will be over budget if the project continues to perform as it has in the past. Based on the fact that this falling behind has been a trend from the beginning, the project will most likely come in behind schedule by around one and a half weeks and over budget by around $3,600. Please see Figure 5.1, which depicts the project's progression in terms of EV, PV, & AC over the life of the project to date.

From this type of chart, a project manager, in a snapshot view, can learn some critical details about the project's performance. Clearly, the AC of the project has nearly always been higher than both PV and EV, meaning that the project, most likely, experienced a negative cost variance. Also, the EV and the PV have been relatively stable, meaning that the project has been on schedule overall; currently,

Table 5.6 Baseline variances 02

WBS item	Baseline duration	Actual duration	Baseline cost	Actual cost
Lot prep and foundation	3 weeks	3.6 weeks	$20,000	$19,500
Framing and masonry	5 weeks	5 weeks	$25,000	$27,500
Rough-ins	3 weeks	4 weeks	$15,000	$17,500
Septic and water	2 weeks	1 weeks	$10,000	$8,500
Drywall, paint, and cabinets	6 weeks	6 weeks	$15,000	$14,000
Interior finishes	4 weeks	2 weeks (i)	$12,000	$4,000
Floor coverings	3 weeks		$6,000	
Decks, guttering, and driveway	2 weeks		$5,000	
Grading and landscaping	2 weeks		$5,000	
i—incomplete	30 weeks	21.6 weeks	$113,000	$91,000

Table 5.7 EVM calculations 02

Status	End week 21
Planned value (PV)	$93,000
Earned value (EV)	$89,000
Actual cost (AC)	$91,000

Table 5.8 EVM results at end of week 21

Cost Variance (CV) =	EV - AC	=	($2,000)
Schedule Variance (SV) =	EV - PV	=	($4,000)
Cost Performance Index (CPI) =	EV / AC	=	0.978
Schedule Performance Index (SPI) =	EV / PV	=	0.957
Estimate to Completion (ETC) =	(BAC - EV) / CP	=	$24,539
Estimate at Completion (EAC) =	AC + ETC	=	$115,539
Alternate ETC =	(BAC - EV) / (CPI * SPI)	=	$25,642
Alternate EAC =	AC + Alt. ETC	=	$116,642

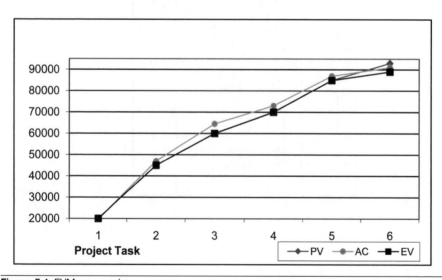

Figure 5.1 EVM progression

however, because the EV is less than the PV, the project is experiencing a schedule variance, which was evidenced in previous calculations. By using a chart such as Figure 5.1, or even a more detailed chart, project managers can quickly see how their project has been performing, and can forecast future project performance.

These types of tables and charts can also be very helpful when preparing to perform a lessons learned evaluation, as they help direct discussions to points in the project where variances existed. By preparing the chart according to varying degrees of detail, the project manager can look at the project from various perspectives, seeking to gain insight during performance analysis. He could also graph the forecasted amounts with actual amounts to determine his forecasting accuracy.

It is my position that EVM has much to offer those within the residential construction industry, even those in smaller firms with fewer formal project management procedures. Not only could EVM be used in construction projects, it could also be used to report performance on internal projects, regardless of type, as well as land development projects. Therefore, those within the industry stand to gain much from this performance management tool.

Risk Management Plan

The risk management plan developed during the planning phase of the construction project is also an integral part of the controlling phase. The risk management plan provides information that is important to the monitoring phase:

1. List of prioritized risk events
2. Risk response strategies

These two items serve as a guide through all parts of the execution phase of the project, as the possibility that these risk events will occur is an ever-present reality. Therefore, the risk management plan should not simply be a document that is filed away; it should be a document used throughout the life of the project. This section discusses how one uses the risk management plan during the controlling phase. There are two steps associated with the risk management plan during the controlling phase of the project lifecycle: review and implement.

Review the Risk Management Plan

If effective planning was done during the planning phase of the project, there may be very little to do during this step in reviewing the plan. However, if not much effort was put into the plan early on, now is the time to revisit that section of the planning phase and complete the risk plan.

What the project manager is primarily looking for are changes or shifts in the dynamics of the project that have not been accounted for in the risk management plan. This could be changes made to the design of the home, the financing of the work, the use of certain subcontractors, or a number of other items. When the project manager develops the risk plan, she is basing it on a number of key assumptions—about who will do the work, what obstacles will arise, and about the various dynamic forces surrounding the project. She develops plans to account for what might come to pass based on those assumptions. If the assumptions change, the plan must be altered to account for that shift in the dynamics of the project. If she fails to do this, the risk plan will not be worth very much, and it might even create more problems.

For the review, each part of the plan should be covered. The project manager should update and reprioritize the risk list. She should review the impact and probability assessments for accuracy based on up-to-date information, and she should review the risk response strategies for feasibility. This review may seem tedious, but if she developed a good plan during the planning phase, the review can be done in a very short amount of time. After reviewing the plan, the project manager may proceed to implementation, which will require some additional work to integrate the plan with the project schedule and other performance evaluation techniques.

Implement the Risk Management Plan

Implementing the risk management plan involves simply putting the systems in place to make certain that the directives of the plan are followed during the construction of the project. The goal is not only to make certain that people are aware of the plan, but they are actively using the plan to monitor and control the execution of the project. It also requires that the plan should be updated and modified as needed as new risks are identified.

Typically, much of the work that is done during the controlling phase of the project is covered in the risk management plan, as a large portion of the controlling phase of the project is focused on managing risks. Many of the tools that the project manager uses during the controlling phase have been identified during the planning phase as valid means of managing the various risks associated with the project. For instance, one of the risks of the project is that there will be cost overruns. In order to help minimize any cost overruns, the risk management plan may include guidelines to monitor purchase orders and accounts payable to ensure that all outgoing payments are properly authorized and that the work has been properly inspected. By having this type of check-and-balance system, the project manager reduces the risk of paying for work that is unacceptable. The risk management plan may also call for the use of earned value management (EVM) techniques, as a means of controlling various aspects of the project's expenditures. The risk management plan can play an integral part in controlling the project during the execution and closing phases of the project.

A few steps are suggested for implementing the risk management plan during the controlling phase of the project; they are:

1. Integrate with project construction schedule
2. Assign tasks
3. Monitor feedback
4. Authorize response strategies
5. Update risk management plan
6. Record lessons learned

These steps are not necessarily sequential; they are better viewed as the various activities engaged in when implementing the risk management plan. This is obviously not to say that the steps might not happen in a sequential manner, but it is not required. The project manager should be flexible in the application of the plan. Now each aspect is discussed in turn.

Integrate with project construction schedule The first step in integrating the risk management plan with the construction project at large is to integrate the various portions of the risk plan to the construction schedule. The construction schedule is focused around scheduling the work to be performed, but it does not typically include the controlling aspects of the projects, which are also necessary for the completion of the project. Some project managers choose to create a few different types of schedules, which can be linked together in most project management software packages.

A project manager may choose to create the following schedules:

- Construction tasks
- Material orders
- Inspection/quality control
- Controlling/monitoring tasks

These may be integrated into one schedule or they may be linked. The method of bringing the information together is not necessarily the focus here. The focus is on understanding that there is much more to a construction project schedule than merely putting the specific construction tasks in some type of sequential order.

One of the most important schedules to create is a schedule for monitoring the project's progress with the various tools and techniques outlined in the risk plan. For instance, at specific points in construction, the schedule will call for updating the EVM analysis for review. Each time this is done, the project manager compares results with the previous results, which will allow him to notice any trends that might be present in the data. Creating and monitoring this schedule is important because it will tell the project manager if any solutions that might have been implemented after the previous reviews are actually improving the performance of the project.

The project manager creates this schedule through reviewing the risk plan and the construction schedule. The specific risk monitoring measures and risk response plans will need to be integrated with the construction schedule in a visible and meaningful way for it to be useful. If the risk plan is scheduled, but the schedule is not followed, the work is for nothing. Therefore, the project manager must hold people accountable to the monitoring scheduling, just as he must hold people accountable to the construction schedule.

Assign tasks After the schedule has been created or while it is being created, the project manager must assign the actual tasks of performing the controlling activities to members of the project team. In some instances, larger construction companies will have a quality control inspector assigned to projects. If this is the case, the majority of the work will be the responsibility of this individual. However, in smaller companies, this is not the case. In smaller companies, often the project manager has sole responsibility for monitoring not only the quality of the work, but also the quality of the processes guiding the work.

Assume, however, for the sake of example, that the project manager has a few team members who will participate with her in implementing the risk plan. If these people were known when the risk plan was initially developed during the planning phase of the project, the assignments have most likely been made. However, this is not necessarily the case, as the dynamics of the project may have shifted somewhat since the plan was initially prepared.

The need may exist to reassign certain portions of the risk plan to various team members. In assigning any task, the project manager must ensure that the individual possesses the skills and knowledge to perform the work. She should review the details of the assignment with each individual to ensure that the person is fully aware of his role and responsibilities. He must also be given the necessary tools and access to the necessary information or software programs to perform the work.

Monitor feedback As the project is progressing and the risk management plan is being implemented according to the schedule that was developed, the team members responsible for monitoring the work will be sending various types of feedback to the project manager for review. This feedback is reviewed to make certain that the project is still within acceptable control limits.

There are a couple of approaches that may be taken when performing inspections. Some project managers might schedule inspections but only request feedback when an issue is identified. This method is used by many people, but it is not recommended. It is best for the project manager to receive some typewritten confirmation that the inspection was performed and that the findings have been recorded, even if all work that has been performed was approved during the inspection. It is even more important for the project manager to receive written feedback when the information will be entered into some type of statistical control system, such as EVM.

Whenever the project manager receives the various feedback reports, he will need to review them for completeness and detail. This does not mean that those preparing the report must write a large amount. Brief reports are actually preferred. But the reports do need to be thorough enough that the project manager can tell that the situation has been properly inspected or reviewed.

At times, the reports will need to be analyzed, or further research may be necessary. The purpose of these types of reports is to trigger the project manager when a problem exists. If the report seems to indicate that a problem is developing or that a risk event has already occurred, the project manager will need to move to the next step of authorizing a response plan. If, however, this is an unexpected risk event, the project manager must research the problem to make certain that the events that led to it are properly understood. Then she can move to implementing a response strategy.

If the feedback is statistical data, such as EVM reports, they must be considered in light of the project as a whole, including any past reports that have been produced. Whenever the project manager is reviewing these types of reports, she is looking for trends. Is the situation improving, is there further decline in the performance of the project, or is all going as planned? These reports do not necessarily provide causes or solutions; they merely indicate that a problem exists. It is up to the project manager to develop solutions and then authorize their implementation, which is the next step of the process.

Authorize response strategies The purpose of monitoring the risks and problems that arise is that response strategies can be implemented as soon as possible. Earlier in the book, the various response strategies were discussed. During the planning phase, response strategies were also prepared for all the known risk events that were considered serious enough to pose a reasonable threat to some aspect of the project. Most often, these plans are prepared by a project manager who is hoping that the plans will never need to be implemented. But when the time comes, in most situations, a feasible plan has been prepared and is ready for implementation.

It is not always the case that the response plan has been prepared, or it may be necessary to modify a previously prepared response plan so that the plan takes into account the current dynamics of the construction project. The steps that have been discussed in previous sections of this chapter will serve the project manager well in preparing any new response plans that might be necessary. In light of the previously stated information, the project manager should keep in mind the following points.

First, the response plan should be as simple as possible. Many times, project managers want to create complex solutions to complex problems. The best response plan is a simple one. Second, the project manager should attempt to make certain that the response deals as directly as possible with the root cause, not simply the symptoms. The root cause analysis method previously discussed will serve as a guide to discovering the root cause at issue. Third, the project manager must project how the response plan will impact other areas of the project and adjust the project as needed to deal with these new factors. This means she should update the schedule, as well as any other areas affected by the response plan. Last, she must monitor the implementation of the response plan to make certain that it has the desired effect. Too often a plan is set in motion and no one verifies that it works. So the plan should include a monitoring of the response plan.

Update risk management plan Throughout this entire process, the project manager must update the risk management plan with the various types and forms of information being collected. The purpose of this is not to create a mountain of paperwork, but to provide a paper trail of how the project has progressed. This is needed for two reasons. First, it is good practice to keep track of key decisions so that they might be analyzed and reviewed at a later date. Second, the paper trail will provide a record of how the builder has addressed different situations in case a lawsuit, insurance claim, or other such issues arise.

Record lessons learned The last step is to record lessons learned. A project manager and project team is continually learning new lessons on each project. Most of the time, individuals will mentally store away a note about a particular situation or process for future reference, which is great. To put an event to memory so that one deals with similar situations in the future in a better manner is a key to personal and professional success.

But it is also important to record that information for others. Recording lessons learned is a key part of the controlling process, which shall be discussed in greater detail later on. For now, however, the project manager must realize the importance of recording any significant events or lessons that transpire during the work of the project. At the completion of the project, a final lessons document should be prepared to act as a summary of the project as a whole.

Specific Performance Evaluation Techniques

Cost Solutions

When a project begins costing more than anticipated, the project manager must investigate the source of the cost overruns and begin taking steps to correct the problem. The investigative process described earlier will serve the project manager well in locating the problem. The problem might be identified through budget reviews, milestone reviews or by the use of EVM techniques. However the problem came to light and the source located, the project manager now turns to what options she has to better control the costs of the project.

The solutions at her disposal will depend on the contractual constraints she is under. For instance, if she is leading an internal project for which the client is her own company, she will most likely have greater flexibility than if the project is for a specific client, engaged in the building process as an active stakeholder. So her options available at any given time will depend on the dynamics and constraints of the project.

Locate new vendors One option is to locate new vendors for materials. For instance, if the project manager was intending to order the majority of materials through one supplier for convenience sake, he may consider seeking out other vendors who have more competitive pricing on different products. For instance, the company for which I work typically purchases the framing package from one company, the trim package from another company, and roofing materials through yet another company. The various companies are more competitive on the different products. Some might believe that the time taken to research different suppliers is not worth the savings, but that would be foolish. Depending on the size of the project, my company realizes a couple of percent savings per project. Although this may only be a few thousand dollars per project, the savings is quite significant when repeated over numerous projects.

The project manager must be careful, though, when using this cost saving strategy, especially if she is dealing with labor subcontractors. She must be careful to not lower the quality of the project to an unacceptable point in order to control costs. Some materials are less expensive because they are of lower quality; and some labor is less expensive because it is of a lower quality, as well.

Modifying plans Another strategy that is often used is to modify the project plan to reduce the complexity or the scope of the work. The project manager will be constrained to varying degrees, depending on the nature of the project. If he is working for a client on a cost-plus contract, the client may choose to modify the plan. If the project manager is working under a fixed-cost contract, it is unlikely that the client will allow the builder to modify the plan because of cost overruns. In this arrangement, cost overruns do not affect the client. Typically, the project manager will have the most flexibility when building a spec home, as the company who employs him is the client.

When modifying the plans of a home, however, the project manager must be careful not to harm the integrity of the home. Some changes, such as using a lower quality of flooring or cabinets can greatly hamper the sale of the property, if it is a spec home. So the project manager must think through any changes. Typically, no one change will greatly reduce the cost of the project. Instead, the project manager makes a few changes, which will add up to more significant savings.

Renegotiating with subcontractors A third option is to renegotiate contracts with subcontractors. Typically, this involves reducing some aspect of the work, but not always. If the project is in peril, the subcontractor may be willing to reduce the price. The theory is that some work is better than no work. However, this method cannot be used during a steady construction market; it can usually only be employed during a major downturn in the local construction market.

Requesting additional funding If it appears that the project is going to run over budget, and the project manager is unable to find a way to reduce the cost of construction to bring the contract back within the budget, he may need to request additional funding for the project. This is typically the last method to employ, as it can have a devastating effect on the project and the project manager's career.

If additional funding is needed due to poor management, the project manager might even be asked to step aside from the project, or he may be placed in some type of probationary status. If the cost increases are valid due to authorized project changes or circumstances beyond the project manager's control, the request for additional funding is not seen as problematic, but necessary to perform the additional work.

Before a project manager requests additional funds, it is imperative that he provide a clear statement as to why the additional funding is necessary. Having a clear and concise reason for the additional funding is more advantageous than are evasive answers that shift blame to other parties.

Schedule Solutions

When the project begins to fall behind schedule, the project manager must launch a course of action that will restore the original timetable as soon as possible. This is not always an easy or even a realistic option. Sometimes, the project manager has to request a time extension, whether due to change orders by the customer or poor management. Whatever the cause, the project manager is responsible for correcting any problems. Here are a few strategies that the project manager can employ to correct the schedule.

Fast tracking Fast tracking is a technique used to accelerate the construction schedule. If a project manager were to review the construction schedule, she would see a fairly linear progression of steps. When one person finishes a task, the next person starts a task, then the next, and so on. In fast tracking the project manager accelerates the project schedule by overlapping steps. For instance, when the time comes to finish the plumbing and the electrical, instead of letting the plumber finish before the electrician comes, the project manager schedules the plumber and the electrician for the same time period. They may get in one another's way a little, but often they can work on opposite sides of the home and pass each other in the middle. This method will speed up the construction schedule, but it can also create problems. The pressured schedule can cause tensions to rise, especially if there is conflict between the subcontractors who are working concurrently. It can also lead to a decreased quality level because of the compressed timetable.

Crashing Crashing occurs when the project manager focuses all available resources on the project's critical path. Because the critical path represents the longest path through the schedule, the project manager puts all the resources toward it; reducing the critical path means reducing the duration of the project.

Crashing is an aptly named technique, because doing it can really crash a project. There are a couple of considerations in crashing a project. As the project manager continues to throw more and more resources at the critical path, she runs the risk of actually changing the critical path. For instance, if the project manager focuses only on the critical path without focusing on tasks that are not part of the critical path, she may finish the tasks on the critical path without completing the other tasks, thus creating a new critical path. For instance, if a project manager focuses on finishing the interior work of the project, setting aside any exterior work, the interior work will be completed, but the exterior work will then need to be done. Originally, the critical path was the interior work, but when it is done, the exterior work becomes the critical path of the project.

Another concern with the crashing technique is the effect that it can have on the cost and the quality of the project. If more workers are hired to perform the work more quickly, there will be additional costs, which can quickly lead to cost overruns. This push to finish the work as quickly as possible can also lead to quality issues, as workers will begin to focus less on the quality of their work and more on the time required to complete it. Therefore, if the project manager decides to crash a project schedule, he should be reconciled to increased costs and lower quality of work in most cases.

Time extension The time may come in the project when a time extension is necessary. If this is so, the project manager should inform the key stakeholders about this as soon as possible. In preparing the request for the time extension, the project manager should outline the following:

- The reason for the delay in completion
- Steps taken to attempt to return to original deadline
- Additional time requested
- Additional costs due to delay
- New schedule identifying key milestones

Presenting this information clearly in the report, the project manager provides the key stakeholders with the data necessary to approve or decline the time extension. The more complete the report, the more likely it is that the time extension will be approved without problems. If the report is purposefully vague, it will invite more questions, which could create further delays. It is best to provide an open and direct account of the situation.

Quality Solutions

During any project, the quality of the work may be called into question. In the best case scenario, the project manager scrutinizes the project, questions the quality, and makes corrections before any other stakeholder identifies the issue. Of course, it doesn't always happen this way. Many times, it is the client or an inspector who calls the work into question.

When the quality of the work is in question, the project manager must inspect the work with the key stakeholders and determine what solutions might exist. Sometimes, the problem is due to some code issue that was missed and can be easily corrected. The work is corrected and then approved by the building inspector. Typically, the project manager expects to have a few items fail inspection during the project, and it is typically not a major issue.

The client might raise the issue of quality. In this situation, the client might not have understood what the finished product would look like. This is an unfortunate situation, but it can happen. For instance, the client may request that a rear concrete patio be poured off a rear door. This decision may be made before the construction has begun, and the client may have a difficult time visualizing the elevations of the finished product. When the work is complete, the client might not like what the finished product actually looks like. In this case, the project manager has met the terms of the contract—he has provided a patio according to the terms. The client may have to simply live with the results or pay to have it modified in some way. Or it might be that the workers have not finished the work according to the terms of the agreement. In this case, the work must be modified so that it is in compliance with the project plan. The project manager is responsible for determining the problem and implementing the solutions that he develops.

Repairs Repair work refers to work to correct something that is deficient in some small way. An example of this would be the vast majority of issues that are listed by the building inspector on an inspection report. Most of the time, the building inspector lists small repairs that the subcontractors will make. Most of the time, the work is not flawed in any major way. Clients are also a source of repair work. This might include an issue with the paint, the flooring, or any other aspect of the construction project. Typically, it can be corrected relatively easily and inexpensively.

Rework Rework is the next level up from repairs. A repair is patching a leak under the kitchen sink due to a faulty fitting. Rework is removing the bathtub and installing a full-size shower due to a mistake by the plumbing subcontractor. Rework refers to major modifications that must be made when the work performed is not in compliance with the project plan. Repairs are seen as a rather routine part of the construction project, whereas rework is not routine and is typically costly.

Having to perform rework can create havoc on both the project's schedule and the project's budget. It should be strenuously avoided. This is the purpose behind the many inspections by the project manager in executing the project. If rework is due to a change order by the client, the project manager will receive additional funding and additional time. However, if it is due to a mistake in construction, the project manager must develop a plan that has the least impact on the project schedule and budget.

While developing the risk management portion of the project plan, the project manager should highlight any unique areas of the project where mistakes would be more likely and more costly. These areas should be flagged and closely monitored to ensure that the work is done correctly to avoid any possible rework.

Improve quality control methods As work on the project progresses, the project manager continually monitors the quality of the project. For different portions of the project, different tools are used. Checklists, material inspections, work inspections, EVM techniques, baseline analysis, and a number of other methods might be employed. The project manager must consider how each of these methods actually helps maintain and improve the quality of the work. He focuses on helpful methods and modifies or abandons unhelpful methods. By remaining open to new methods and improvement to current methods, the project manager can continually improve the quality of the project.

Interpersonal Solutions

Too often, a project manager will overlook the interpersonal issues just below the surface of the project. Failing to deal with these issues can be just as devastating as neglecting problems in other areas. A disgruntled employee or poorly managed subcontractor can harm other employees or the project itself. The project manager should protect herself, her company, the project, her workers, and her client by managing the interpersonal aspects of the project.

Phases of a team To manage the interpersonal aspects well, the project manager must be attune to the phases or lifecycle of a team. In 1965, Bruce Tuckman originally introduced the four phases of team development, which became known as Tuckman's Stages.[1] The phases are:

1. Forming
2. Storming
3. Norming
4. Performing
5. Adjourning (fifth phase added in 1977)

Some teams will deal with different phases in different ways depending on the dynamics of each group, but every team passes through these five phases. Understanding the phases will help the project manager know what to expect as the project begins, and equip him to deal with the individual members of the team.

Forming

In the forming stage of team development, the team is assembled from a group of individuals who have been assigned the task of working together on a common goal. Sometimes the people will know each other, but maybe not. Many times, the project manager's team will be people from the construction company with whom he has worked in the past. If this is the case, the team may move quickly through the first two phases, but this is not necessarily so—especially if the project manager or some of the team members are new.

During the forming stage, the project manager makes certain that everyone is introduced and that everyone understands the other person's role in the project. If the project manager sets expectations and boundaries for the team members, there should be less fighting and posturing for various assignments and responsibilities.

---

[1]Alasdair White. "From Comfort Zone to Performance Management" Article can be accessed at www.pm-solutions.com

Storming
In this stage, the team members begin settling in to their roles and positions in the project. Several might attempt to settle into the same role, which creates conflict. This is especially true when new people are working together, and they are trying to impress the project manager. They may bicker over assignments and the like. Some project managers provide very few directions during this stage, allowing the team to work out their problems. This can be very dangerous for the project. It is my opinion that the better job the project manager does with forming the team through the assignment of roles and responsibilities, the less conflict he will have to manage during the second phase of team development.

Regardless of the project manager's efforts toward clearly outlining roles and responsibilities, he should expect some conflict. Conflict among people is an inherent part of the human condition. The difficult part is to ensure that the conflict is not destructive to the project, and that it ends as soon as possible.

Norming
During the norming phase, the team members have come to know one another better and have settled into their respective roles within the project. This is not to say that conflict will not arise, but it should be the exception to the norm. This, of course, is dependent on the culture surrounding the team. If one works in an environment that is very high pressure and around certain personality types, conflict may be an unavoidable part of the project. For instance, if the project has a hard deadline that the team is having difficulty meeting, conflict may continue throughout the life of the project. If, however, the environment is less demanding, such as it can be in much residential construction, one would expect little conflict during this stage.

Performing
In the performing phase the team reaches a high level of performance capability. The members of the team understand their roles, and they are working diligently within them toward the common goal of completing the project.

I am currently working on a project team which recently passed through these phases. The company recently sold a spec home that was nearing completion in a second-home community. The home was ready for flooring, to be followed by finishes from the plumbing, electrical, HVAC subcontractors, and the like. The buyer elected to modify the home on a number of levels that included tearing out finished walls and making rather major modifications to various aspects of the home. Once the client had clarified all the changes that she desired, the various subcontractors and other key stakeholders were brought together to work through how the work should proceed in the most efficient manner. Even though the individuals have worked together for years, they were brought together in a unique setting, and one could see them move through the various stages of team development. First, they established the roles they should play; they argued over which bits should be done first and what was the best plan; they normalized relations by accepting various concessions to each person's individual plan; then they began the work.

Adjourning

The final phase is the adjourning phase, in which the need for the team has ended, and each of the team members moves on to a new assignment. During the adjourning phase, the project manager is concerned with closing the project and transitioning to a new project. The aspects of this phase will be discussed more fully in the next chapter on closing the project.

Conflict resolution With this understanding of team development, the project manager can see a specific phase in which conflict among the team is expected. This in no way means that conflict is limited to that portion of team development. Conflict will be a continual part of the project, as the project brings competing people together in ambiguous and volatile situations. Some may think that this is a dim view of the construction project, but it is a realistic view. The project manager is a mediator among various stakeholders who have different interests and goals in mind when they come together.

If the project manager seeks to completely avoid conflict, he is playing the part of a fool. Avoiding conflict only leads to growing unrest; it is best to deal with conflict directly, but in a manner that does not create further conflict. Conflict resolution guides offer a number of strategies. Such texts use key words such as *compromise* or phrases such as *win-win* or *win-lose*, and they offer helpful information. However, in my opinion there is a certain amount of conflict that is to be expected in every project; the question is how does one reduce and manage the conflict?

In the world of financial management, the various financial investment vehicles (stocks, bonds, mutual funds, CDs, etc.) one can utilize have a degree of risk associated with it. There is the risk that inflation will rise, devaluing the investment (inflation risk). There is the risk that one's original investment will lose its value (principal risk). There is also the risk that the investment cannot easily be converted to cash (liquidity risk). There are other types of risk as well. Each of these risks derives from either one or a variety of market forces, and the investor must deal with these risks if she is going to invest in the market. The goal of most investment strategies is to remove the risks as much as possible so as to not only protect the original investment, but to earn an acceptable return on investment. For instance, one might purchase oil futures as a hedge against inflation and the falling value of the dollar, a technique employed in mid-2008 by a number of firms causing a dramatic increase in the cost of gasoline in the United States. Whatever strategies are taken, there is still a measure of risk in every investment.

The same is true with conflict in projects. There is a measurable amount of conflict associated with every project. One can never entirely remove the risk of conflict from a construction project, but one can most certainly take steps to mitigate it. In my opinion, there are two primary factors that have a high degree of influence on the amount of conflict present in a project: communication and expectations.

Communication

Communication is the first factor that has a strong influence on the conflict that is present in any project. Where there is poorly managed communication, there is a higher degree of conflict. Poor communication fosters distrust and uncertainty, which give rise to conflict. When a stakeholder feels cut off from the information that she needs, she will take the steps necessary to learn the information, and typically she takes an aggressive posturing to receive the most attention. Many times, this is one-sided conflict. Most of the time, the project manager is not purposefully withholding information, he is just doing a poor job of communicating the information that the client or other stakeholder desires to know. He might be surprised when the conflict arises, not realizing there were any problems. This in return creates a defensive posturing, which leads to escalating aggression between the two parties. Clear and open lines of communication will do much to remedy the conflict that can arise in a construction project. However, clear lines of communication alone will not solve the problem. Bad news communicated clearly can create as much conflict as can good news communicated poorly.

Expectations

Each stakeholder has expectations in relation to the project. The construction company management believes that the project will be a profitable venture; the project manager believes that it will be an opportunity to expand his skills and advance his career; the client believes that she will be receiving a well-built home at a good value; the subcontractors believe that they will be properly compensated for their work. The list goes on; everyone who is working on the project has either conscious or subconscious expectations about the project. The project manager is responsible for managing these expectations. This means that he is not only responsible for communicating expectations, but for shaping realistic expectations. A project manager who fails to do this is likely overseeing a project that is riddled with conflict.

The first major factor discussed was communication, which obviously concerns making certain that the expectations are properly communicated. The second major factor is primarily concerned with shaping or molding expectations. In each project, as stated above, various stakeholders come into the project with various expectations. It is the responsibility of the project manager to ensure that the expectations, in general, conform to the project plan. Consider an obvious example.

The client purchases a spec-built home that calls for a fireplace. At the time of purchase, the fireplace has not yet been installed, but the contract states that a fireplace will be installed in the home before completion. The client is looking forward to using the fireplace. She remembers sitting by the fire as a child during those cold winter months in Michigan throughout the holiday season at her grandparents' home. She remembers the seeming simplicity of life; the good conversation and the feeling of contentedness and joy among the family, as the aroma of hickory wood subtly permeated these joyful memories. She is thinking of recreating those same types of memories with her grandchildren in this new home, which is supposed to be done by the holiday season when the family will come visit. It is clear that the buyer is expecting a wood-burning fireplace; but the builder is planning to install gas logs, as the chimney is not a wood-burning chimney, but merely a facet of the design to create a cabin feel in the home.

When the buyer learns that the fireplace is not wood burning but a gas log insert, the project manager will have to endure some conflict. The reason is not because the builder purposefully misled, but because the builder did not clearly communicate what the buyer's expectations should be. The contract simply stated fireplace, and the builder assumed that any person would recognize that the design was for a gas log insert and not for a wood-burning fireplace. This is the type of conflict that can lead to a buyer's cancelling a contract, which can in turn lead to a lawsuit. Although this is a fabricated example, it is not unrealistic. There are hundreds of opportunities during the construction project for the various stakeholders to misunderstand, which leads to their having either unclear or unrealistic expectations. The project manager needs to take special care to educate, not just the client, but the other stakeholders as to what they should expect. When the project progresses as the project manager said it would, the chance of conflict is greatly minimized. This is especially true when custom design items, such as fireplace mantles, flooring designs, or custom cabinetry, are called for. The project manager must take special care to ensure that the client clearly understands what to expect. If she does not think that she will like the results, she needs to either alter her plan or abandon the issue altogether.

By making certain that the client and the other various stakeholders have clear and realistic expectations, the project manager can greatly reduce conflict during the construction project. However, if the project manager does not do this well, then managing conflict will simply be a part of his job.

Now that some of the various techniques for correcting issues have been addressed, let's turn toward a method of process improvement known as the theory of constraints.

Method of Process Improvement

The purpose of all process improvement techniques is to do work more efficiently and effectively. What hinders efficient and effective work are the problems that arise from any number of sources. With process improvement, the project manager is not so much dealing with specific problems with the technical aspects of work, but with how the work is managed.

One of the challenges of process improvement in construction management is the need for a system of improvement that can work within the unique setting of construction projects. I believe that the application of the theory of constraints (TOC) to the construction environment is just such a system.

Theory of Constraints

The TOC offers five simple steps for improving project performance. This discussion will in no way begin to cover either the breadth or depth of the TOC, but it will provide a very understandable portion that can be immediately implemented in any project.

The project manager must follow five basic steps in order to use TOC. The titles of the five steps to follow are borrowed from a book by William Dettmer (1997) entitled *Goldratt's Theory of Constraints: A Systems Approach to Continuous Improvement*. Although the titles of each phase are from Dettmer's book, the information to follow the title is original to this book. Those five steps are:

1. Identify the constraint(s)
2. Decide how to exploit the constraint(s)
3. Subordinate everything else to the above decision
4. Elevate the constraint(s)
5. Rework, or find the next constraint(s), then return to step one

Identify the constraint(s) A constraint is anything that is hindering the project from being completed according to the project plan. In any given project, there are a variety of constraints. With process improvement, the primary focus is on the process that has been put in place to manage and perform the work. The project manager is focusing on the methods of management currently being used that could be improved. Whatever holds back better performance is a constraint. This means that what the constraint is now, may not be the constraint later. As the work progresses, different constraints will arise, which must be dealt with. Therefore, what is identified as the constraint early in the project may not be the same thing constraining the project at the midway point. Identifying the constraint is an iterative process.

Exploit the constraint(s) Exploiting the constraint refers to removing the constraint. The TOC does not specify a specific means of doing this necessarily, as each project has its own challenges. But the point here is that whatever is causing the problem is corrected. Once a practice has been identified as constraining the project, the project manager must modify or develop new methods to exploit or remove the constraint. This requires creativity and problem solving skills.

Subordinate to the decision This means to coordinate other decisions with the decision made to exploit the constraint. The point here is that when the project manager decides to correct a problem, the changes needed to correct the problem will typically create a ripple effect through the project, which must be dealt with. If the ripple effect is ignored, he may create a more difficult situation than the one he was attempting to solve. For instance, let us say that a certain constraint was identified—late delivery by a certain supplier. Exploiting the constraint would entail developing a system of dealing with this supplier that removed the problem. For instance, the project manager could give the supplier a few more lead days or could locate a new supplier. Either way, the change made to correct the problem must be implemented. This is what it means to subordinate to the decision. If the project manager decided to use a new supplier, all interested parties would need to be informed to not order from the old supplier. They would need to order from the new supplier specified by the project manager. This is what it means to subordinate to the decision.

Elevate the constraint(s) This refers to the need to elevate or to inspect the original constraint that was identified, now that the solution has been implemented. If the solution implemented was the correct one to solve the problem, when the project manager considers the constraint, she will find that it is no longer the primary constraint. If the constraint is still as problematic as before, the solution was ineffective, or not correctly implemented.

Rework Rework refers to performing any modification necessary to address the constraint if the original solution did not solve the problem. This step also refers to the need to rework through the process yet again. This is a method of continuous improvement, which means that once the project manager corrects the current primary problem, she still needs to find what has now become the next primary constraint. In this way, the process is iterative; she continuously moves through the steps throughout the life of the project as a means of continuing to improve project performance.

Now that the various aspects of performance evaluation have been discussed, let's examine the second major focus of the controlling phase: integrated change control.

INTEGRATED CHANGE CONTROL

Integrated change control is one of the most difficult aspects of both executing and controlling a project. Although it is covered in the controlling section of this book, it is very much a part of the execution phase of the project. In general, integrated change control is concerned with controlling changes to the work of the project and the project plan. The source of these changes can come from a variety of sources, which is discussed shortly.

Integrated change control involves two aspects of the construction project in this book. Technically, *integrated change control* refers to a system that processes how changes are made to the project plan. This includes changes to the actual construction of the home. It also refers to how one changes the project baselines or alters the risk response plan or changes vendors or accounting systems or any other change to the way the project was planned. So while there is most assuredly a concern about how the actual construction work is modified, it also is concerned with changes in how the project is managed.

This book focuses on two broad areas of change: (1) process change control and (2) construction change control. Process change control refers to the process of modifying how the project is managed. Construction change control refers to the process of implementing construction change order requests submitted during the construction of the home. However, before these two areas of integrated change control are covered, let's look at a general definition, which is applicable to the changes both to the construction of the home and to the construction project plan. After this, each of the types of systems is discussed individually.

Integrated Change Control Defined

Integrated change control refers to the process of recognizing, reviewing, authorizing, implementing, and recording all change requests associated with either the project plan or the project deliverables. If the system fails in any one of the five aspects mentioned above, problems can result that can harm the project.

Recognizing

The need for, or the existence of, a change must be recognized before it can be properly managed. In any given project, changes might occur that are not recognized by the project team at large. In this instance, a team member, subcontractor, or other stakeholder might modify a portion of the deliverables or the project plan without informing anyone else of the change. This can start a chain of events that creates problems for the success of the project. Or it might cause no noticeable change during the life of the project, but may create a flaw in the deliverables that is not seen for some time to come. If a change is not recognized, it can probably not be moved through the integrated change control system, which can lead to unrealized and negative effects.

A change is recognized when it is brought not only to the attention of an individual associated with the project, but when it is brought to bear, as it were, on the project at large, meaning that those who need to know do know. The source of the change may be from a variety of sources. It could come from a customer, who wants to modify a portion of the construction plan. It could come from the building inspector, who states that a portion of the project work, as planned, is in violation of a portion of the building code. It could also come from a subcontractor, who is unable to perform the work as he had agreed for some reason. The point is through either a voluntary or an involuntary means, a change must be made.

The client should not simply be able to call a subcontractor and make a change, nor should the building inspector be able to direct work without letting the project manager know, so that he can deal with the modifications. There must be a formal means of recognizing the change and then moving it through the proper processes to make certain that the project plan is adjusted. Typically, this is done through a written change request.

If the client requests that the work be modified in some way or another, the changes she desires must be written down and presented to the project manager. Many times, the client simply calls the project manager or another project team member and vocalizes what her desires are. If this is the case, the project manager should put it in writing and submit it for review. If the statement accurately reflects her desires, it should be signed by her, at which point it can progress to the next stage.

If, however, the change is necessary because an issue has arisen with the building inspector or a subcontractor, the project manager is typically provided with an inspection report by the building inspector, outlining the necessary modifications.

Regardless of the source, the change is recorded in writing to ensure that both parties are in agreement as to what the change requested actually is. At times, this may require the client or building inspector to draw some type of diagram as a guide or illustration. The goal of getting the change request in writing is to have a guide that the project manager can follow through the remaining phases of the change control system.

Reviewing

Once the written change request has been submitted, a review process begins. Each change request is scrutinized on a number of levels, regardless of the source of the request. Questions to ask about every change request are:

- What is the cost?
- How much time will it take?

- How will it affect the project's budget?
- How will it affect the project's schedule?
- What inspection implications are there (building codes)?
- How will it affect the quality of the project as a whole?
- What materials will be necessary?
- Which subcontractors will need to be involved?
- Are there alternative solutions?
- Is funding available as part of the construction loan?

These are a sampling of the types of questions that the project manager will be asking whenever he begins reviewing the change request. These questions need to be asked regardless of the source of the change request. Most people would think that these types of questions need to be asked whenever the client submits a change request, but they should also be asked whenever the building inspector submits a notice which results in a change request.

Answering these questions is no mean task. It takes time to gather all the information. Sometimes seemingly minor change requests can have tremendous impact. For instance, the simple removal of the pantry in a kitchen can have a big impact. When one goes to remove those walls, he will have to deal with every system that may be inside the walls: plumbing, electrical, HVAC. Each subcontractor must return to the job site to remove work and redo it. Depending on the stage of construction, this may require tearing out floors and subflooring, cutting drywall, removing cabinets, and a host of other actions. It will require work to be redone and reinspected. The implications of a seemingly simple change can be quite staggering to someone who is unfamiliar with construction work. Clients usually have an especially difficult time understanding why some changes cost so much.

In reviewing the change request, the project manager must carefully think through each part of the change request. He should speak with each worker who will be involved in the change request, making certain that both the worker and the project manager understand the extent of the work required to implement the change.

An impact study will also need to be prepared, which shows the client and other stakeholders what impact this change order will have on the project. The additional cost is only one factor to be considered. The additional time required for the work and the additional inspections needed must also be considered. The project manager must ensure that the client understands how the change order may affect the completion date of the entire project. If the project manager does not take the time to perform this type of analysis, the client will be making an uninformed decision, which may lead to unnecessary issues later in the project.

Part of the review process is developing a cost estimate for the change order. Many times, builders underestimate the cost of change orders by only taking the raw cost of the change order into account. This is a foolish mistake. Typically, change orders create a higher degree of risk in regards to the client's satisfaction. It creates additional problems in a number of areas, and the builder should be compensated for this additional work. The project manager may want to develop a markup rate that is over and above the actual cost of implementing the change order if the change order is requested by the buyer. This is not to gouge the buyer, but it is to fully compensate the builder for the extra work.

Authorizing

After the change order has been reviewed and presented to the client and other key stakeholders, it must be authorized before it can be implemented. Depending on the type of change in question, it may be necessary to get the authorization of the client (construction change order) or the authorization of the project manager or project sponsor (process change order). The client must authorize change orders, but they are not the only ones. The project manager must as well.

In most construction contracts, there is a clause that gives the client the right to request change orders that are considered to be within the general scope of the project. The language is somewhat vague in most contracts, and vague language is typically only made clear through a lawsuit. The project manager must tread very carefully. I am not an attorney, so this section does not constitute legal advice, but it is based on my experience.

Sometimes the client wants a change request that the project manager either knows will be more work than it will be worth or believes that the client will ultimately be unsatisfied with the results. Other times the project manager simply does not want to do a requested change order. In this case, the project manager should have a clear understanding of the construction company's policy on such matters. His employer might say that it does not matter what the project manager wants. Or the construction company may offer a degree of discretion to the project manager to refuse some change orders.

If the project manager decides that he does not want to do a change order, his best move is to attempt to talk the client out of the request. The reasons that a project manager might not want to perform a specific change order can be numerous. For instance, the change request may be beyond the project manager's knowledge area. Or the project manager may not simply want to perform a change order because it would lengthen the project's duration to a point that would interfere with the project he has planned for the future. It is important to remember that the construction contract will need to be referenced before the project manager refuses to perform a change request. Often times, the construction contract allows the client to make certain types of change requests. If the type of change requested is allowable according to the contract, the project manager must do it if he is unable to persuade the client to rescind the request.

Persuading the client to drop a change request can usually be done by explaining the reasons that the project manager does not want to do the work. This tactic works best when the project manager believes that the client would not truly be satisfied with the quality or aesthetics of the end result. If possible, he should provide examples or samples of the product as evidence and clearly explain his concerns. If he is unable to convince the client to rescind her change request, he must either flatly refuse the change request, or he must concede and perform the work. If he chooses to refuse to implement the change request, he must be ready for the confrontation that will follow, which could possibly lead to a lawsuit. Therefore, the project manager is cautioned against taking this path too quickly or without strong reasons.

If, however, the project manager is willing to authorize the change order, the client must simply authorize the work as well. When the project manager presents the findings of the review of the change order to the client, it is best to guide the client through the proposal. This will ensure that the client understands the full effects of the change order on the project as a whole. Many times, he is surprised by the cost, the time necessary, or the impact on the remainder of the project. Once the surprise wears off, it can sometimes turn to aggression, depending on the temperament of the client. The project manager is cautioned to be prepared for this type of response.

If the client expresses surprise, the project manager can simply walk the client through the impact study showing how the change order will affect the project as a whole. If the presentation is clear, the client understands the enormous amount of work that is often required for seemingly small changes. Then the client can either approve the change order or cancel it, once he understands the true impact. Upon approval, the change order moves to the implementing phase. If the change order is cancelled, it will move to the recording phase.

Implementing

Once the change order has been authorized by the project manager, client, and another key stakeholder, it can be implemented. Implementing change orders can have widespread effects on the project as a whole. Depending on the stage of construction and the extent of the change request, all work on certain aspects of the project might be halted. The project manager must redirect workers and modify the project plan in a number of areas.

First, the project manager will address the impact on the progression of the work. The affected workers must be informed about the change request, and changes to the schedule must be made. Materials are ordered before the work can progress, and new inspections are scheduled. The subcontractors or other workers involved in the change request must receive any updated drawings or specifications that will provide them with the information they need to perform the change request.

Too often, project managers rush into change orders too quickly, which can result in mistakes and create further delays and increased costs. The project manager must move decisively, but not necessarily too hurriedly. It is better to take the time to do the work right the first time, than rush and risk creating problems.

While the materials are being ordered and workers are being rescheduled, changes are being made to the project plan as well. The project manager updates the project schedule to reflect the change order. This requires not only adding some tasks in the schedule, but communicating those changes to all necessary stakeholders so that they are aware of changes to the schedule.

The manager must also alter the schedule baseline, cost baseline, and quality baseline as well. Statistical measurements of project performance, such as EVM, are based on comparing actual progress to planned progress (baselines). If the plan changes, the baseline must also change to reflect all the planned work.

There are situations in which the baselines should not be altered. If the change order is required because a portion of the work failed inspection and a major part has to be redone, changes to the baselines should probably not be made. When the project manager makes a mistake that sets the project back in time or creates a cost overrun, she should not have the freedom to alter the baselines so that it appears that no mistake has been made. The monitoring statistics should reflect the fact that she made a mistake and that it is creating problems. But when a client voluntarily makes a change, the project manager should not be penalized because the project will take additional time and cost additional resources.

Once the project manager updates the project plan to reflect the modifications required by the change request on the project plan and the actual home construction specifications, she must begin monitoring the change request and the other work on the project as before. If the change request has unanticipated effects on the project, such as creating further delays and additional costs, she should notify key stakeholders as to those unanticipated effects.

Recording

The last aspect of the integrated change control system is recording. The facts and information surrounding change requests must be accurately recorded for future reference. Typically, a higher percentage of contract disputes arise from change orders than from the original terms of the construction project. Many change order requests are an attempt to correct what the client perceives as a problem. Often what the client believes will correct the perceived deficiency does not in fact solve the problem, but only creates additional problems. Another reason that change orders often lead to conflict is that they present a greater risk of resulting in unintended consequences. These unintended consequences can not only affect the client, but also the project manager.

How many times has a project manager agreed to a change request too quickly, not realizing the implications of making the changes? If a project manager accepts a change order too quickly without considering the implications, both the client and the project manager's superior may become upset, and the project may be harmed due to cost increases or time delays. The project manager may attempt to pass the blame to the client or another party, but, ultimately, it is he who caused the problem through his foolish and rash decision.

The project manager will be held accountable; he will be held to a higher standard than anyone else. This is why I emphasize a formal process for handling even small change requests. Small change orders that are managed poorly have a way of becoming large problems that cannot be managed well.

Recording information and receiving acceptance and confirmation throughout the process is critical. Recording means that all the associated documents should be kept on file. It also means that the project manager should receive written acceptance of the change order after it has been implemented. In this way, the project manager has taken as many steps as he can to ensure that there is a minimal backlash in the event that a problem arises in the future.

Now that a working definition of what a change control system involves, the two different types of change control systems are discussed briefly, with special attention given to system design, system implementation, and system monitoring. The construction change order system is discussed first, and then the process change order system is briefly covered.

Construction Change Control

As stated previously, the construction change control system is specifically concerned with managing the construction change order requests submitted by any stakeholder to the project manager for consideration. The construction change control system involves developing a means of accomplishing all the tasks discussed in the preceding section.

Construction Change Control System Design

With the previous section in mind, the project manager should move to prepare an integrated change control system. The design of a construction change control system is not a difficult process. In designing the system, the project manager is seeking the best way to move a change request through the stages mentioned earlier for her specific project.

Many times, the project manager will not have to design a system herself. A sample flowchart of such a system is shown in Figure 5.2. The company for which

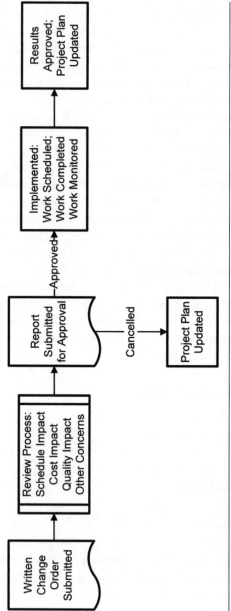

Figure 5.2 Change control diagram

she works will have already worked out the procedures that it would like followed on the company's construction project, and the project manager must simply implement that system in her current project. This is one of the benefits of working for a larger construction firm, which has the advantage of having project management systems in place. But if the project manager works for a small company, or if she is the company, she will need to develop a system for handling the various types of change orders that she might encounter during a construction project.

When creating a change control system, the project manager has the option of creating different classifications of change orders that will treat the various change orders differently, or she can treat all change orders the same. It depends on what she believes will be useful. She will want to think through a few factors when designing a system. Those factors are: complexity, cost, source, and stakeholders.

Complexity Complexity is both a consideration and a type of classification. A project manager who has been involved for any time in construction projects knows that the complexity of a change order is a major consideration. The more complex the change order, the larger the ripple effect through the entire project. For instance, if the client asks the project manager to have a few extra outlets installed in a den early in construction before the rough-ins are complete, the project manager knows this is not going to be a very complex change order. It is a rather cut-and-dried affair. But if the client wants a bathroom ripped out and the size of the adjoining bedroom increased after the drywall has been installed, the project manager has a rather major change order on his hands.

As a general rule, the more complex the change order the more expensive, time consuming, and potentially risky the work is. Therefore, when designing a system to deal with change orders, the project manager must establish how the more complex change orders will be treated. What type of vetting process will they be subjected to before approval? Who is responsible for making the decision to either implement it or not? These are the types of things one must consider.

Cost Cost in this situation refers to both quantitative and qualitative costs. Quantitative costs include both monetary resources and time. The cost of a change order should be considered in both real dollars spent and time taken to implement the change order. Is there a point at which the project manager will begin to refuse to do change orders because of the extensions in time it is taking? Is there a point at which the home will cost so much that the buyer will not be able to secure financing? There is also a nonquantitative cost that cannot easily be measured. Change orders not only add more time to the project and more cost, they also add pressure and frustration to the project.

A good change order system might not necessarily lower the cost of a change order, but it can lessen the time required to research and implement the change order, and it can most assuredly reduce the frustrations that accompany so many change order requests. So here the project manager should be asking himself if he wants to accept the high costs associated with not having a change order system. Not having a system means looking for pieces of paper or napkins on which change order requests were written. It means not remembering exactly what the customer wanted because the project manager believed he would be able to remember when the time came. It means underpricing the cost of the change order and having to explain to the project sponsor why the profit margin for the project is lower than anticipated. The cost of not having an integrated change control system is just one of the additional costs that one must add to change orders.

Developing a good system that manages and minimizes costs, both qualitative and quantitative, creates a much higher degree of satisfaction not only for the client but also for the project manager. When a change order request is submitted, the project manager will have a better chance of truly estimating all the costs if he has considered these factors when designing a change order system.

Source The source of a change order is a factor that affects how the change order is handled. At some point in construction, a client might come up with a creative change order that had just not previously occurred to her. She will present it in this manner: "I do not know if it would be possible, but if you could do this, that or the other, then I will really appreciate it." She realizes that it might be a long shot, but she would be genuinely thankful if the change order could be implemented. In this instance, the project manager does not treat this change order request as less important than one where the client says that a particular change order is a must have, but he does realize that if the cost is too great or the impact too much, it is unlikely that she will ever have to implement the wishful change order.

If a change order is submitted by an engineer or a building inspector in the form of a failed inspection report, the project manager can be sure that he will be implementing the change order request, most likely at his company's expense. These types of change order are usually nonnegotiable. The home must conform to the building code and acceptable building practices. If the home is deficient, the project manager will be forced to correct the deficiency.

So there are both mandatory and nonmandatory change order requests. This does not mean that the change orders are treated differently, but the project manager realizes that one must be implemented, whereas the other might only possibly be implemented. If the change order is the result of a failed inspection, the project manager will not likely receive additional funding or additional time to correct the issue. There will be pressure on her to speed up the project in some way to make up for any lost time. Whereas, if the client requests a change order, he will accept any schedule or cost increases due to his change request. Therefore, the project manager may develop a different track for addressing each of the change request types.

Stakeholders The last factor to consider is what stakeholders must be brought into the system for any given change order. For instance, if the change order is due to a failed inspection, the client, inspectors, project manager, and subcontractor will most likely be involved. If it is a simple request about changing the type of wood flooring before the floor has been installed, the client, project manager, and flooring subcontractor will be involved. Whereas some change orders may require any number of workers to return, depending on the complexity of the request. The client's banker may even be required to approve the work, as the change request may necessitate additional funding on the construction loan. The project manager is responsible for determining whom the change order will affect and how each of them will be involved in researching, approving, and then implementing the change order.

After these and any other factors the project manager can think of have been considered, he is ready to design his integrated change control system. This may appear to be a very complex process, but this is not the case. The process should be as simple as possible, but it should also be thorough. Change orders tend to create problems when people do not properly research and inform. Developing a system ensures that each change order is thoroughly considered before being authorized. The flow of a change control system, in general, is as shown in Figure 5.2, but the project manager will need to work out the specific details of how these factors can best be implemented for a specific project.

Implementing the Construction Change Control System

Implementing a construction change control system primarily involves two steps. First the system must be clearly communicated. This does not simply mean that people are told that such a system exists, but how the system works must also be explained. Second, the system must be utilized. This may seem like an obvious step, but I can hardly recount all the systems that have been designed and developed only to never actually be used.

Communicating The fact that the system exists must be communicated, as well as how the process works. A diagram similar to the one shown in Figure 5.2 may be helpful in explaining how the change order system will work. The system must be explained to the primary team members and to certain stakeholders, such as the client, subcontractors, material vendors, and any other relevant stakeholder.

The reason that the system needs to be communicated so broadly is so that people will understand that a process is in place and that they are expected to follow it. Too often the client comes to the job site and sees the electrician working, begins asking questions, and then gets an idea for something he wants changed. As the electrician is here in front of him, he asks the electrician to add or modify something. Without a clearly communicated and understood change order process in place, the electrician might make the change without providing a cost estimate or getting approval from the project manager and then submits a bill for the extra work. When the bill is presented to the client, he is likely to be completely taken aback. He had no idea that it would cost $400 to run the power supply for the hot tub he is thinking about buying after he moves in. After all, it was only a simple power outlet. Of course, a simple power outlet will not operate a large hot tub, but the client does not know that, and he was not sufficiently informed ahead of time. He may refuse to pay, which means that the project manager must pay the bill or the electrician might place a mechanic's lien against the property for unpaid work. This scenario occurs again and again on job sites when the subcontractors and the clients get together and create changes without following the system that is in place to protect both the subcontractor and the client from misunderstandings. When the system is explained, the client knows not to ask the electrician to make a change without first consulting the project manager, and the electrician knows not to perform any work that he does not have written authorization from the project manager to perform.

Utilizing Not only must the construction change control system be communicated to the various project stakeholders, but it must actually be utilized. Many companies have these types of systems on paper, but they do not seem to exist on the job site. They are not used because it takes time to write things down and do research and get approval. The workers would prefer to guesstimate a price and get to work. This is a recipe for not only conflict, but lawsuits and judgments.

The project manager is responsible for ensuring that the home is built according to the plans and specifications agreed upon. She is also responsible for making certain that the project is managed according to the project plan. If she fails to do this, she is placing the stakeholders at risk. Using the change control system will take time and effort, and it is not always fun or enjoyable, but it is essential.

Monitoring the Construction Change Control System

So far the system has been defined, designed, and implemented, but this is not all that is necessary. The system must also be monitored. Systems are subject to two basic failures: They can be broken and they can be misunderstood. A broken system simply does not work. The people follow the guidelines, but following the guidelines does not yield the right results. At some point, the system breaks down and creates problems when it was designed to solve or avoid them. If systems are misunderstood, people won't follow them properly. In this situation, someone tries to bypass a portion of the system or she fails to understand how the system actually works. People who fail to understand the system need to be corrected and trained. Those who are trying to bypass the system need to be questioned. The project manager should attempt to understand why they are bypassing a particular portion of the system. If it is because the system is broken, that will need to be corrected. If it is because they are lazy or incompetent, they may need to be reassigned to the unemployment line.

There are at least two aspects to monitoring the system: audits and feedback. An audit is when someone takes the time to review how things are passing through the system. Is the system accomplishing the purpose that it was created to accomplish? Audits can take place at various intervals or when a problem is reported. The timing will depend on the judgment of the project manager. Soliciting feedback from stakeholders is also a means of monitoring the construction control system. Those who are actually using the system can provide the best feedback about whether the system is accomplishing the goal it was designed for or how it might be improved. The project manager might solicit feedback when team members encounter a problem or have a suggestion throughout the project. However, he might also want to want to create a survey that can be sent out during the closing phase of the project creating the opportunity for the stakeholders to provide targeted feedback about this and other areas of the project. This information can then be used to correct problems and improve the system.

Now that the construction change control system has been covered, a brief overview will be provided about the process change control system.

Process Change Control

Construction change control is focused on those change orders that affect the construction-specific aspects of the home; process change control is focused on managing the more internal change requests, which concern how the project is managed. For instance, if the client requests that a half bath be added in the basement of the home, this is a construction change order. If the project sponsor asks that the project manager begin using a critical path method to manage the construction schedule rather than the critical chain method, this is a process-related change request.

Process-related change requests are more internal in nature than construction change orders. Although the client might not be aware of the process-related change requests, she would be highly involved in construction-related change requests. This does not mean that the process change control is less important because it does not typically involve the client; it plays a critical role in project success.

The factors associated with the process change control system are very similar in nature to that of the construction change control system. Each of the systems can be defined as having the same basic steps, which were covered previously in the section, and are:

1. Recognizing
2. Reviewing
3. Authorizing
4. Implementing
5. Recording

These five steps or phases should be included in the design aspects of any process change control system. Whereas in the construction change control system, the client is the primary stakeholder, in the process change control system, the project manager and the project sponsor are typically considered the primary stakeholders or decision makers.

Process Change Control System Design

In the design of a construction change control system, four areas or factors are key:

- Complexity
- Cost
- Source
- Stakeholders

These four factors, which are a part of any type of change order request, play an important role in how much time is invested in a particular change order. In addition to these four, one additional factor is added to the process change control system: impact. The impact that the change order will have on the way that the project is managed is of great concern. This is not to say that impact is not an important factor in construction change control systems, but those considerations are most likely accounted for within the complexity factor. Here special attention is given to the impact that a process change control can have on the project. Sometimes seemingly small and insignificant decisions can have a profound impact on the success of the project. The project manager must be certain that she is considering those ramifications when looking at a potential process change control request.

These five factors should be the determining factors in how a change order is processed. An efficient process change control system recognizes that not all change requests are equal. Different members of the project team should have the authority to approve change requests that are within their level of authority and skill. A management system that is inflexible can create barriers to project success. In designing a system, the project manager should train the members of the management team to consider change requests through the lens of these factors, and then the requests should be sent through the appropriate channels for approval and implementation. Now each of the five factors shall be considered briefly in turn.

Complexity Complexity refers to the overall involvement required to implement the proposed change order. Complexity could refer to the involvement required to change the project plan in order to implement the change, or it could refer to the difficulty of performing the work that is outlined by the change order request. Remember that process change requests are focused on internal management issues. Consider that a change request is submitted to use green paper to print notices rather than yellow paper, which is standard, due to the savings that can be achieved by switching to green paper. This is an incredibly simple change request. There is no complexity involved here. Running this type of change request through an integrated process change control system is a waste of time and money.

If the builder were to request that the project manager move from using a critical path method for developing the project schedule to a critical chain method, this is a change that would greatly alter the project schedule and a host of other parts of the project and potentially confuse and confound those working on it. This would be no mean change request, and it should only be done if required by the project sponsor. In this case, it would be worth the time and energy to fully vet such a change order request.

In designing a process change control system, the project manager should run all change requests through the same basic process, but he should probably flag those more complex projects in such a way that particular attention is given to them. This is not to say that complex change requests should necessarily be given more priority or focus than less complex projects, but more complex projects need to be considered carefully because of how they will possibly affect the other factors, such as cost and impact.

Cost A second factor to consider when designing a change request system is the emphasis or the focus that should be given to those change requests that have a higher cost factor. Any experienced project manager would want to spend some time considering those change requests that proposed great costs, whether in time or real money. Often a cost-benefit analysis should be performed to consider the cost compared to the potential or proposed benefit of the change order request.

Cost is, however, only one of the factors that should be considered. Just because a change order has a low cost with supposed high benefits, the benefits may not be that great when other aspects are considered. Therefore, cost alone is not and should not be a decisive factor. When a project manager chooses solely based on the short-term cost, he runs the risk of reducing the quality level of the project. Therefore, other factors need to be considered alongside the cost factor.

Source This is one of the most critical factors of a process change control request. This is primarily the case because of the pitfalls associated with it. Typically, one will either overrate the source or will underrate the source to the detriment of the project. Consider the following example:

A new vice president is hired at the construction company's corporate office and is responsible for improving project performance through new and innovative methods. Her first day on the job, she begins sending out directives for how projects will now be managed: the method of budget development, schedule development, and the like. This is not only to apply to future projects, but to current projects as well. The reader, a project manager for the company, is in the middle of a highly involved project, which is on schedule and within budget and appears to be heading toward the completion date without any foreseeable problems arising. This new directive, however, requires him to update his project's schedule, budget, and other affected areas of the project plan. After reviewing the proposed changes, the project manager determines that the new method of managing the project will take a great deal of time and effort to implement on his current project. Although it will help him with future projects it will not improve his current project. If anything, there will be additional costs and time to implement the new project management system on his current project. Based on the factors given, what should the project manager do?

Even if the new system is a better way to manage projects in general, implementing it at this point in this project would be foolish. It would in no way benefit the actual work of the project, and it would most likely increase both the cost of the project and the time required to complete the project because it would require reworking how the project is managed, not necessarily how the project is performed. The project manager should propose that the changes not be made, given these factors, but he may be required to implement them regardless. In this case, the source is overrated. Because she is a new vice president, her decisions stand. Project managers are sometimes required to implement changes that will not actually improve the project. In this situation, they should register their objections, and then do the work their employer instructs them to do.

Another problem is that often times a source is underrated, and the project suffers. For instance, a junior member of the project team or even an intern might suggest some change to the way that the projects are managed, which, if actually implemented, would prove to be a good suggestion. However, because of the pride of the project manager, the request is never given the opportunity to be proven as a good idea.

A change order system must take the source into consideration, but a source should not be the final factor influencing a decision. The benefit the change request will make to the project should be a hinge point of any process change control request.

Stakeholders Just as with construction change control requests, process change control requests will affect stakeholders and this is an important consideration. The more stakeholders involved and affected, the larger the ripple is likely to be in the project. In designing a system, the project manager should consider how the information is going to be communicated to the various stakeholders. This is the primary design consideration. The project plan includes a communication plan that lists stakeholders, as well as means of communicating with those stakeholders. This will be an important part of any change control system design.

Impact Impact is a factor that was not discussed in the construction change control section but is included here. This is not because construction change orders do not have an impact on the project. Impact can often be overlooked when considering process change control requests; internal change requests can at first glance seem to be minor and insignificant. However, sometimes small decisions have a lasting impact on a project. For instance, choosing not to perform a minor inspection because it seems unnecessary could lead to missing something that could create problems later on. The impact should be analyzed, and the potential impact should determine how a change order moves through the system with more attention given to those that have a higher impact.

Implementing a Process Change Control System

The implementation of a process change control system is not that different from that of the construction change control system. The construction system implementation focused on two items: first, communicating the process and instructions to involved parties; and second, utilizing the system. These are also the key steps for implementing the process-related system. The project team members must be told what the system is. They must have access to the forms necessary to write up a request or the person to contact to submit a request. They must also be instructed to utilize the system that has been designed. Many will be inclined to simply ask permission from the project manager to do something at variance with the project plan for some reason or another. Even if the request seems like a good idea, it should still be submitted to the system so that future projects will have the benefit of the improvement suggested by this particular team member. It will also serve to inform others of the modification to the project plan. The development of a process flow chart will be helpful in showing and reminding people of the proper process.

Monitoring the Process Change Control System

As with any system, controls must be in place to ensure that the system is being appropriately followed and utilized. Too often, a system is designed and followed initially only to be abandoned at some future point for some reason or another. The monitoring of the systems will indicate a few things.

First, monitoring the system will make certain that it is utilized. If people know that the system will be audited at regular intervals, they are more likely to use the processes in place rather than abandoning them. This will ensure that the system is followed, and it will also serve to bring any deviations to the forefront so the project manager can take appropriate action.

Second, monitoring the system will allow the project manager to determine if there are problems that need to be fixed or improvements that can be made. If the system is abandoned because it is creating more problems than it is solving, it needs to be abandoned, but it should be done so openly and knowingly so that the system can be fixed. The primary purpose of the change control systems is to make certain that changes are properly implemented and to make certain that all affected parties are notified. If the system is abandoned, the changes might not be fully considered, and not everyone will be made aware of the changes. This can create confusion and threaten project success. The process control system, just like the construction control system, should utilize both audits and feedback, as discussed previously as a means of monitoring the system.

This chapter has focused on the fourth phase of the project lifecycle— controlling the construction project. Controlling the project involves two primary focuses: managing performance and managing change. This phase runs concurrent with the execution phase of the project, and it ends when the project ends. The last phase of the project lifecycle is the closing phase, which is the topic of the next chapter.

REFERENCES

Barlow, M. J., and T. A. Klingelhoets. 1997. *Earned value supports enterprise-wide project management.* Chicago: Project Management Institute 28th Annual Seminars & Symposium.

Budd, C. I., and C. S. Budd. 2005. *A practical guide to earned value project management.* Vienna, VA: Management Concepts.

Dettmer, William H. 1997. *Goldratt's theory of constraints: A systems approach to continuous improvement.* Milwaukee, WI: ASQ Quality Press.

Fleming, Q. W., and J. M. Koppelman. 1999, October. *The earned value body of knowledge.* Philadelphia: Proceedings of the 30th Annual Project Management Institute 1999 Seminars & Symposium.

Hillson, D. 2004. *Earned value management and risk management: A practical synergy.* Anaheim, CA: PMI 2004 Global Congress Proceedings.

Project Management Institute. 2008. *A Guide to the Project Management Body of Knowledge,* 4th ed. Newtown Square, PA: Project Management Institute.

6

CLOSING THE CONSTRUCTION PROJECT

The closing phase is the last phase of the construction project lifecycle. First the project was initiated, then it was planned, then it was executed, while it was executed it was controlled, and now the time has come to close the project. The emotions surrounding the closing of a project are typically a mixture of excitement and relief. Excitement is generated by the successful completion of this project and the hope of a new project. Relief is experienced when everything comes together in the end. This, of course, assumes that the project being closed was a successful one. The emotions involved in closing an unsuccessful project can range from tedium to hostility. This is a special case, which shall be discussed separately later in the chapter.

In general, closing the project involves assuring completion of every aspect of the project, not only for the project manager but also for the various stakeholders of the project. Everyone has his or her own list of completion requirements before the project will be closed. The closing procedures primarily focus on the tasks involving the project manager and the client but will also possibly include subcontractors, vendors, team members, financial institutions, attorneys, surveyors, engineers, building inspectors, and any other stakeholders. During the planning phase, the project manager developed a preliminary plan for closing the project, but that plan must be revisited and updated with any new information that might have come to light during the execution of the project.

Closing procedures cover a wide range of both topics and stakeholders. At minimum, closing procedures will cover the following areas:

- Client
- Job site
- Construction stakeholders—vendors, subcontractors, inspectors, etc.
- Project file
- Project team
- Audit procedures

By thinking through the closing phase categorically as presented above, the project manager creates a higher probability of ensuring that all relevant issues are covered and closed during this phase. Each of these categories is covered here; at the end a special section will cover some issues related to closing an unsuccessful project.

CLIENT

Most project managers are quite familiar with procedures and issues involved in closing the aspects of the project directly related to the client. The focus on the client-related issues is appropriate, as these issues are the most critical aspect of the closing phase. If the project manager can successfully close with the client, the rest of the closing is typically merely a matter of paperwork. This is not to say that other areas are unimportant, but to stress the importance of closing well with the client.

Typically, when *closing* and *construction* are uttered in the same sentence, one thinks of the closing that takes place at the attorney's office or at the bank, when title is transferred to the client, and the seller or builder receives payment for the property. Although this is an important and integral part of the closing phase, it is only one part. There is much work that must be done to get to this point, and there is some work to be done after this point. Because, however, the closing transaction is such an integral part of the closing proceedings related to the client, it will be treated as a hinge or reference point for this section of the chapter. The information will be presented in a linear manner, starting with some of the steps that are part of the construction WBS, which lead into these final closing procedures.

As can be seen from previewing Figure 6.1, there are a few steps that lead to the official closing with the client, and there are a couple of steps that follow.

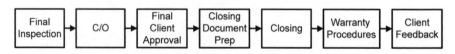

Figure 6.1 Client related closing procedures

Final Inspection

The final inspection is performed by the local building inspection office to ensure that all work that has been completed on the home is in compliance with the applicable building codes and zoning ordinances. Depending on the area and guidelines applicable, this may be a one-time inspection, or it may require a couple of phases. Regardless of the process, it is required before the builder can acquire a Certificate of Occupancy, which is the authorization from the local regulatory board stating that the home can now be occupied by a resident.

Issuance of Certificate of Occupancy

The certificate of occupancy (C/O) is one of the most important documents that the builder acquires, because it states that the home meets the applicable building and zoning codes. With this document, the client can have the power turned on, the water turned on, and obtain insurance with a homeowner's insurance policy.

Simply because the project manager has acquired this document, however, the client is not required to accept the project as complete. The contract between the client and project manager might include work other than that which is covered by the building inspector. For instance, many building inspection offices do not require the flooring to be installed in the home for a C/O to be issued. It is doubtful that many clients would accept a home as done when the flooring (carpet, hardwoods, tile, etc.) has not been installed. Because there can be a different scale of acceptance between what the building inspector will accept and what the client will accept, receiving a C/O is not enough. The project manager must also have the final approval of the client.

Final Client Approval

After the final inspection is completed, the WBS calls for a series of inspections to be performed by the client or the client's agent. Often times, the client will hire an independent home inspector, or an architect associated with the project, or an engineering firm to perform the final inspection with the client. Typically, these inspections focus on more cosmetic issues than structural or mechanical, but the latter may also be included.

Receiving the final approval of the client may take a couple of inspections and reinspections. Regardless of the process, the project manager must receive final approval if he hopes to close on the project. Sometimes, legitimate disagreements will arise as to whether some aspect of the work is in compliance with the construction contract. When these types of disputes arise, the only party that can interpret what exactly the contract states is the courts. Arbitration may work, but arbitration simply tries to get the parties to reach an agreement through varying degrees of compromise. Only the courts can provide a definitive determination as to what is and is not in compliance with the terms of the construction contract. Knowing this, the project manager should proceed cautiously. Up to this point, she should have been having inspections and reviews with the client throughout the entire construction process. This final inspection should not give rise to any major new issue but should simply be a process of receiving final approval. If, however, an issue of higher priority arises, the project manager will have to determine whether to modify the aspect of the home to fit the client's desires or to hold her ground and face a possible lawsuit. This is not a decision to be considered lightly. Most often, it is easier to bite the proverbial bullet and deal with the problem.

After the client has given final approval, the project manager is almost ready to meet the client at the closing table in order to sign all the final documents and receive final payment. Typically, the closing date will have been set by the client and the client's financier at a time beyond the final inspection, to allow time for the punch list items to be addressed. During this interim period, the builder gathers certain documents that are required for the closing.

Builder-provided Closing Documents and Final Financial Statement

During construction, the project manager compiles a project file filled with various documents. Included in this file are various reports and records, which will be pertinent and necessary for the closing. Items such as the soil inspection report, any engineering reports, surveys, inspection reports, wood-destroying insect treatment reports, lien-waivers, and any other documents that were obtained during construction are included. Depending on the nature of the closing, these and other documents may be required by the banking institution providing financing for the property, as a means of assuring that the work was properly completed.

The builder may also be required to complete a variety of forms, depending on the loan type. If the loan is through the Federal Housing Administration (FHA) or Veterans Affairs (VA), that agency will require various forms to be completed and submitted as part of the loan approval and closing process. If a project manager is unfamiliar with any of the forms, the lending officer can typically provide instructions. These documents must be completed by closing, in most cases. Another document that needs to be prepared is the final financial statement for the project. Invariably, the client will have made changes during construction, which will most likely lead to an increase in the cost of the project. Depending on the structure and terms of the contract, these items might or might not have been paid for at the time they were ordered. If payment has not been received, the project manager prepares a bill to be paid either by the proceeds from the loan or by the client. If there are overages, the project manager ensures that all costs are included and that documentation is available to support those extra charges. It is also important to include a copy of the signed change order agreement, showing that the client authorized the work to be performed.

In rare cases, the client will be due a reduction in the amount requested in the final payment, or even a refund. This typically happens when a client does not entirely consume an allowance that was allocated as part of the contract. For instance, if the contract called for flooring expenses in the amount of $15,000, but the client only spent $14,000, then she will be due a refund of $1,000 if she has not consumed the credit elsewhere through a change order of some type.

In preparing this financial statement, the project manager must be both detailed and clear. The client should be able to read the document without much trouble; it should not be so detailed as to make it overly confusing. Typically, the financial statement is in summary form, and the project manager will be able to provide supporting documentation if a request is made.

Once the documents required for the closing have been compiled, the project manager or the representative from the construction company is ready to attend closing to receive the final payment for the project.

Property Closing

Construction projects and construction contracts come in all types and forms. Because of this, the role the builder takes at closing can vary. In some cases, the builder does not even attend closing, but he will receive a final payment as a result of it. Other times, the builder might be present as a seller, who has sold a spec home to the client. In other cases, closing may simply constitute the builder's receiving a final payment from a client who hired him to build a home on the client's property. Because of the different roles and positions in which the builder can be, there is no one flow that the process always follows.

There is, however, one thing that the builder is always seeking, regardless of the actual closing process—final payment. The builder might have been receiving payments throughout construction, or very few to no payments. At closing, the builder always expects to receive a check. The builder hopes that the amount received exceeds the amount that he needs to pay out to other parties.

In addition, it is a good idea for the builder to provide a packet to the client that includes copies of all the documents discussed in the previous section. This way, the client will not call in a few years looking for a document that the builder has long ago archived.

Warranty Procedures

Either as part of the closing proceedings or after the closing proceedings, the project manager finalizes with the client what the company's policies are for utilizing the construction warranty that the builder provides. After the client moves into the home, she will find items of varying degrees of importance that were missed in the inspections. These items might include doors that do not close properly, cabinet shelves that are not level, or a piece of flooring that creaks. I suggest that the project manager ask the client to live in the home for a few weeks and keep track of any of these minor issues in some type of log. Then she can submit the entire list. In this way, the project manager can address these small items in one or two visits, with the least amount of cost to the company.

If, however, any issues arise that are of more importance than a squeaking step, the client should call the project manager as soon as possible. This type of item would be a leak of some type, an electrical issue, an HVAC issue, or a drainage issue that is causing water to enter the crawl space or basement. These items need to be addressed as soon as possible, as they may only get worse with time, and create worse problems.

In general, warranty requests should be sent in writing, either by E-mail, fax, or letter, and the builder should log them in a special warranty file associated with each project. Acknowledgement of receipt should be sent to the client and logged, and the appropriate repair person should be scheduled. Once the work is completed, approval of the repair should be acknowledged and recorded. A company who manages its warranty well creates the opportunity for future referrals.

Client Feedback

The final aspect of this process is client feedback. Most of the time when a person does not ask a question, it is because he is afraid of the answer. People, in general, think that they can avoid negative outcomes by simply not asking the question. This head-in-the-sand mentality is fatal to the long-term success of a company. Gaining honest feedback from the client on the various aspects of the project is a vital part of improving performance and customer satisfaction. A survey should be developed and sent to the client, as well as some type of gift for completing the survey, such as a gift card to a home improvement store or a local restaurant.

The feedback received must be analyzed and discussed among members of the project team. It can even be integrated into the performance evaluations, which are discussed later in the chapter. Both positive and negative feedback provide the opportunity for the company to improve its performance and correct any past mistakes. It should be viewed as a positive event that is vital for the health of the organization.

The steps presented here should be seen as one stream of the closing process. There are other streams of activity that are also active during this time. A graphical depiction is shown in Figure 6.2.

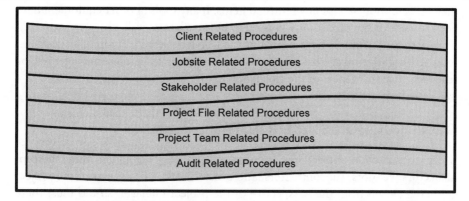

Figure 6.2 Aspects of closing phase

As one can see from in Figure 6.2, the closing procedures related directly to the client's closing on the property is just one of the many aspects of the closing phase. Too often, the project manager becomes transfixed on that phase to the detriment of the others. There is no doubt that it does have a higher level of importance than some of the other phases. However, to elevate it above its proper place within the scheme of the closing phase is to risk completing one of the other phases inappropriately.

JOB SITE

The centrality of the job site through the construction project means that a number of items will be stored and misplaced there. After the final inspection has been completed, and the client has approved any items that were noted in the inspection reports, the job site must be transformed to a home for the client. A thorough cleanup is performed first. The interior and exterior of the home are combed through to remove any remaining items associated with construction. This way when the client arrives, she will be ready to begin moving into the property.

Depending on the project, there may be some unused materials that must be returned. Some materials, such as excess flooring, may have been purchased by the client, and she will make a decision about what to do. Sometimes, the client will keep leftover tile, or hardwood flooring, or even shingles in case a repair is necessary in the future. Whatever materials are not left on the job site are either transferred to another job or returned for store credit. Whatever is done with the unused materials, the accounts must be reconciled with the client to make certain that she is reimbursed for any returned items. If any tools or other equipment have been rented for the project, it will need to be returned as well.

After the materials have been returned or transferred, equipment has been returned, and the job site has received a final cleaning, the project manager will be ready to close this aspect of the closing phase.

CONSTRUCTION STAKEHOLDERS

Every construction project has a diverse group of stakeholders ranging from inspectors, vendors, subcontractors, and others. Just as the aspects of the project are being closed from the perspective of the project manager, so too must these various stakeholders close the project from their perspective. The project manager reviews the relationship that each stakeholder has had to the project and completes any outstanding issues.

In general, there are four areas of focus in closing out issues related to construction stakeholders:

- Work contracts
- Accounts payable
- Lien waivers
- Performance reviews

Closing work contracts simply involves reviewing the contracts related to each stakeholder and ensuring that each party has met the contractual obligations. Without question, the stakeholder will have already performed the work contracted for, but the contract may call for additional services, such as a followup inspection report or a warranty document. Whatever the terms may be, the project manager is responsible for ensuring that both parties have met the terms of the contract before closing out the project. If a stakeholder has not met his obligation, the project manager will be responsible for taking steps necessary to ensure that the terms are met.

As part of the contract review, the project manager reviews the accounts payable to make certain that all payments have been made to those who participated in the work of the project. Typically, this can be done by simply running a report, but it is also wise to make certain that a receipt is included in the stakeholder's file by which he has acknowledged payment. Any outstanding balances must be resolved so that the project manager may accomplish the next step.

The third item the project manager must acquire is a signed lien waiver from all subcontractors stating that they have received payment in full for any and all work performed on the project. Many times this is required by the client's bank as part of the loan closing proceedings. Whether or not the client's loan documentation requires it, this is considered a best business practice for the project manager.

If a dispute were to arise in the future, the project manager has the lien waivers on file as a means to deal with any false claims.

Some project managers forget to address workers that the client has asked to work on the job site during construction. Although it is unlikely that the client has hired a separate construction crew, it is not uncommon for the client to hire individuals, such as a surveyor or home inspector, to provide their services. The project manager must make certain that these individuals have been paid as well, to avoid any future confusion. It is not that the project manager's company would be required to pay for those services, but by acquiring the lien waiver, the company has the documentation in hand to manage any future claim.

The last item related to construction stakeholders is performance reviews. Often project managers do not think to provide performance reviews for stakeholders, because these individuals or companies are not direct employees. It is unlikely that the stakeholder would receive criticism well. However, this should not hinder the project manager from doing performance reviews, even if they are only used internally.

These performance reviews should rate performance based on professionalism, skill level, scheduling, flexibility, cost, and any other areas the project manager believes to be pertinent. If possible, he should review these with the stakeholder (primarily subcontractors), especially if the stakeholder's performance on this job puts any work on future jobs in jeopardy. Primarily, however, these reviews should be filed with the company and reviewed before hiring the subcontractor to work on a future project. In this way, future projects have the experience of the current project to draw on. For a construction company that wants to develop a reliable base of vendors and subcontractors, this will prove to be a very helpful tool.

PROJECT FILE

Project file is a broad term that can cover any number of documents associated with the project: baselines, planning documents, work and purchase orders, and work contracts. In this part of the closing phase, the project file is closed out and prepared as a future reference document. The first step is to review the information and update any documents to reflect actual results. For instance, if the project schedule or cost baseline or quality baseline did not receive a final update, now is the time. This way when someone looks back at the project, she will see information that reflects the actual results of the project, not the results that materialized at various points in the project. If the project manager fails to make these final updates to the project plan, he risks basing future projects on inaccurate information. Future decision makers will rely on the record of this project, and they should feel confident that they are looking at accurate information.

A second step to closing the project file is completing any project lessons learned reports. In general, a lesson learned is exactly what it sounds like—a lesson that the project manager or project team learned during the project. It might, for instance, refer to a lesson learned in planning the project or executing the project. It might relate to a certain type of material or building technique, or a lesson associated with a certain vendor or subcontractor.

The project manager may choose one of a few methods to compile lessons learned. Some use a brainstorming session in which the project team is brought in to discuss the variety of lessons learned, and someone takes down the final version. Or the project manager might simply pass out forms that various team members complete and turn back in at a specified time. Considering the time constraints that may be on a project manager and project team, the second option may be the best. If it is used, then the project manager should have the documents compiled and sent back to all of the team members as a packet so that they might be able to benefit from each others' observations.

The final step in closing out the project file is to prepare it for archiving. This typically involves two steps. First, it involves making certain that all the parts of the project file are present in final form and properly ordered. If this is done, the person who might reference the project file in the future will have access to all the information that was available to the project manager at the time of the project.

Second, the project manager creates a summary and table of contents for the project file. This allows the contents to be quickly referenced. The summary should include highlights of the project, such as the parties involved, the type of home built, the construction time, the budget, as well as a comparison chart showing the performance of the project. The table of contents acts as a guide showing the reader what documents are available in the project file, so that he will not waste time seeking something that is not present to begin with. On a secondary note, the person performing this job should make certain there is no personal information present in the file that could compromise the identity of a party involved, such as a Social Security number or other personal data. This information should be blacked out or shredded as a safeguard.

Another option that is becoming more popular is to keep a paper file for a couple of years but to create a digital file for long-term storage. The digital file offers a number of advantages. First, it does not require actual storage space in the office of the construction company. Second, if properly backed up, it is much more secure than paper documents sitting in a filing cabinet. Third, if properly indexed, a digital file offers much more flexibility and speed to the one seeking information, as she merely needs to locate the file on the company's server and search for the document she needs to reference. This is a great help when preparing for future projects, as one can simply open parts of the file and update it to reflect the realities of the new project. This is a solution that should be considered seriously by any builder who wants to use past files in a meaningful and productive manner.

PROJECT TEAM

The project team is also one of the aspects of the closing phase that needs special attention. Depending on the size of the organization, the project team may only be the project manager, or it may comprise a number of people performing the various tasks of the project. If the project team is the project manager, this is a fairly easy step. The purpose of this step is to provide performance reviews of the members of the project team and to reassign project team members to new projects.

Conducting performance reviews with project team members is a delicate task. The standard way of doing it is for the project manager to call in the various team members one at a time and give an opinion of their performance. A personal review such as this should use not only the project manager's opinion, however. It is best to have the various team members complete performance evaluation sheets about each of the other project team members, including the project manager. The sheets that are applicable to the project team members should be reviewed by the project manager and used to provide praise for what the person does well and constructive criticism for areas that need improvement.

The performance evaluations that relate to the project manager should be forwarded to the project manager's superior in the company, and the superior should conduct a performance evaluation with the project manager along the same lines as those done with the members of the project team. The purpose is to give an honest assessment of his work in light of what his peers think and have observed throughout the project. If this is done well, it offers an excellent opportunity for not only individual improvement, but also for organizational health and improvement.

The second goal of this aspect of the closing phase is to reassign project team members to new projects. Depending on the size of the construction projects the company is involved in, the project team members may already be working on multiple projects. If this is the case, the transition will be easy; if not, the organization will have to spend some time planning what the team members are to be doing during any interim phase until a new project becomes available.

AUDIT PROCEDURES

While all the previous work is going on, the project manager or an assigned team member must oversee all of it to ensure that it is done properly. Making certain that the project closes well is an important aspect of project success. Typically, the person assigned to oversee the closing is merely working through the closing plan that was developed during the planning phase of the project.

CLOSING AN UNSUCCESSFUL PROJECT

This is not an enjoyable topic to write about, as it is not an enjoyable topic to ponder. There are a hundred sayings regarding failure. Certainly, everyone has heard someone say that success never tastes as good as it can until one has experienced failure. Or they have heard an employer say something to the effect that she would never hire someone who had not been fired because that person will work harder not to go through the experience again. These types of sayings have one thing in common—they were said by someone who has failed. A project manager who has failed at a project can blame whomever he wants, but in the end, he always thinks of what might have been if he had treated some situation differently. I know this because I have been a part of failed projects, and I do not recommend the experience. However, here are a few hints to survive it.

There are different ways of defining an unsuccessful project. The project might be unsuccessful from the perspective that it costs more money, took more time, did not meet the expectations of quality, or something similar. This type of project might be considered a failure, but the project manager can still salvage the project to the point that it closes, and the company collects the payment due. If this is the case, the closing steps already covered in this chapter will work. This section primarily deals with a project that is in process and irreconcilable differences arise, which lead to the need to terminate the project. Three steps are recommended for dealing with this type of project:

1. Consult the company's attorney
2. Communicate carefully
3. Cancel all work

First, the company's attorney should be brought in to provide advice on how to proceed. In this instance, the attorney will provide advice as to whether the company will file a lawsuit or will be the recipient of a lawsuit. In either case, the attorney will lay out a course of action for the project manager to follow, which provides the greatest protection to the company. It is vital that the project manager both communicate and follow these instructions with the utmost concern. Further problems can be created through irresponsible action.

Second, the project manager and the project team must communicate carefully with those on the other side of the situation. Communication should be kept to whatever level the company's attorney suggests. It may be best if the project team refrains entirely from communicating with the opposing parties, allowing all communication to pass either through the company's attorney or the project manager. If contact is made, the details of those conversations should be logged so as to provide a written record.

Last, the project manager must cancel all work in accordance with directives from the company's attorney. If an irreconcilable dispute has actually arisen, the project manager does not want to continue to expend company resources, when there is no reason to hope that they will be reimbursed for the work. Once again, the project manager should follow whatever steps the attorney gives as best as possible.

Working through this type of process is a miserable one, and it should be avoided at all cost. Lessons are learned, but it is better to learn the lessons by reading a book instead of living through it.

This phase has been given the least amount of coverage not because it is not important, but because the steps associated with closing the project are fairly direct and self-explanatory. In general, the project manager is simply working to close off any open issues, which have not previously been addressed. Taking the time to develop a closing plan that incorporates the issues discussed here will prove beneficial not only in the short run, but in the long run, as the company will have access to information to learn from and guide them in the future.

CONCLUDING COMMENTS

When I started writing this book, I sat down with a local contractor who not only has field experience, but also has experience teaching, and I presented my idea for the structure of this book. He, like me, found it to be a helpful way of approaching not only a book on construction, but, more importantly, a helpful way to approach construction projects. I hope that you, the reader, can say the same thing now that we have come to the end. The purpose of this book is to show someone how to integrate the various aspects of construction management through the use of the concept of the project lifecycle, which I think provides a good lens for learning about the various topics related to construction management. It allows someone to see better how the various topics work out in the field. To the extent that this book helps someone understand and perform construction projects, it has accomplished its purpose.

INDEX